MW00577895

Us

Air Almanac

for Marine Navigation

With a Comparison to the Nautical Almanac
and
Extended Discussion of the Sky Diagrams

Copyright © 2019 David Burch
All rights reserved.

UK Copyrights.
Pages F3, F4, A20-A23, A122-A123, A125, A130-A152, A157-A167 of the *Air Almanac* and the front inside cover are protected by international copyright law. All rights reserved. Sections of these pages have been reproduced under the terms of the UK Open Government Licence http://www.nationalarchives.gov.uk/doc/open-government-licence/version/2. The source of these pages is Her Majesty's Nautical Almanac Office, United Kingdom Hydrographic Office.

ISBN 978-0-914025-65-8

Published by

Starpath Publications

3050 NW 63rd Street, Seattle, WA 98107

Manufactured in the United States of America

starpathpublications.com

10 9 8 7 6 5 4 3 2 1

Preface

This book is intended for readers who are either already familiar with celestial navigation or are planning to take up the subject, and want to learn about optional resources. This booklet does not teach celestial navigation, nor is it a copy of the *Air Almanac*.

There are marine navigators who prefer the *Air Almanac* over the *Nautical Almanac*. This booklet compares the two almanacs so others can make their choice as well.

The Sky Diagrams of the *Air Almanac* are an important resource for all navigators, but they can easily be overlooked by those who rely on the *Nautical Almanac* alone. With that in mind, we give these diagrams here a well deserved review and comparison with competing manual methods of predicting the best bodies to use for a round of sextant sights with an eye toward optimizing the accuracy of the resulting position fix. The right choice of bodies is crucial to the final accuracy.

For an introduction to or review of celestial navigation, see *Celestial Navigation: A Complete Home Study Course*.

Table of Contents

1. INTRODUCTION

1.1 Background

Finding position at sea from sextant sights of celestial bodies (sun, moon, stars, and planets) requires astronomical data on the locations of these bodies, essentially every second, throughout the year. Mariners typically obtain such data from an annual publication called the *Nautical Almanac.* This publication is jointly copyrighted by the US and UK governments. The official government printing of this publication is expensive ($52), but one US company is licensed to print and sell the almanac for a lower price, and call it the "Commercial Edition" ($30).

The British Admiralty also publishes a hard cover edition: *Nautical Almanac, NP 314* ($60). The content of this book is identical to the US version.

There are less expensive (even free) alternatives to the content of the *Nautical Almanac,* one of which, the *Air Almanac,* is the subject at hand. This is an attractive alternative for several reasons. It is an official publication from the same agency that produces the *Nautical Almanac*; it contains the same astronomical data needed for navigation; and it is a free publication, available as PDF download.

The main task of this booklet is to show, with specific examples, that all the required data for marine navigation are indeed included in the *Air almanac,* and illustrate how to access the data, in light of the different presentations and correction tables.

Besides the basic celestial navigation data and correction tables needed, which are common to both publications, each has its own unique resources that benefit ocean navigation. There are also differences in the layout of the data, both of which are covered with examples in this booklet. Looking ahead, the *Air Almanac*'s Sky Diagrams, used for best sights prediction, is a feature that mariners might choose to use, even if the *Air Almanac* is not their first choice for other data.

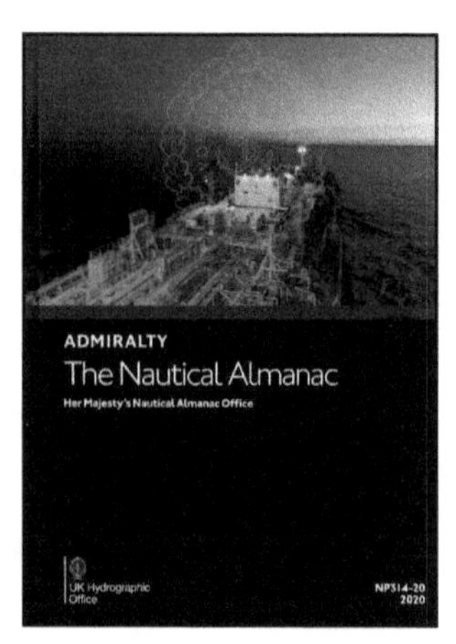

Figure 1-1. *Three versions of the official US-UK Nautical Almanac. The content and layout of the pages, and page numbers are identical in all three.*

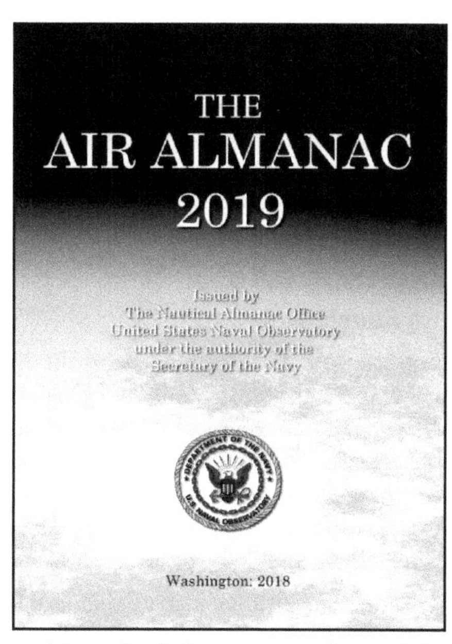

Figure 1-2. *Cover of the USNO* Air Almanac. *This a free document distributed as a PDF file from the US Government Bookstore. The cover image is included in the PDF.*

There is also a British Admiralty publication called *The UK Air Almanac, AP1602*. This publication is also a free annual PDF download, however the only astronomical (astro) data it includes are sun and moon rise and set times, and related resources for estimating daylight. The title is misleading; it is not a contender for air or nautical almanac data.

Another print alternative to the *Nautical Almanac* is *Brown's Nautical Almanac*. This is an annual hardback book from Scotland that sells in the for about $85. It includes *exact replicas* of the *Nautical Almanac* Daily Pages, along with extensive Coast Pilot data and resources, including tides, with special emphasis on the British Isles. Beyond that local information, this is a comprehensive navigation reference book with many unique resources. Even an outdated version would still be a valuable addition to the nav station, but it would not likely be a first choice for astro data alone. This publication dates from 1864. Comparing our own 1945 and 2019 editions shows that the cover has barely changed over that period.

The book called *Reed's 2020 Nautical Almanac* is another British publication, but this one is, these days, misnamed; it no longer includes anything we would call astro or nautical almanac data; it is rather a cross between a coast pilot and cruising guide for the British Isles. Historically, however, this book *with the same title* did include full nautical almanac data, but that Boston-based publication went out of print in the late 1990s. This new version is published in the UK.

Figure 1-3. *Brown's includes full Nautical Almanac data, plus very much more. An excellent resource, especially for those in the sailing mostly in the British Isles.*

Other options for astro data come in the form of computer or mobile device apps that compute the data and corrections. Some even output pages that look similar to those of the *Nautical Almanac*. These products, however, rarely produce all of the other data we expect from an almanac, and they are susceptible to the vulnerabilities of electronics at sea, such as battery charge, saltwater, and rough rides in a small boat in a seaway. These are not considered here as viable alternatives to official printed almanacs.

1.2 Sources

Both the *Air Almanac* (AA) and the *Nautical Almanac* (NA) are produced by the US Naval Observatory's Nautical Almanac Office. The government edition of the *Nautical Almanac* (orange hard cover) is available at the Government Bookstore online, and from selected navigation supply outlets. The commercial edition of the *Nautical Almanac* (blue cover, paperback) is available online at several outlets, as well as numerous navigation supply stores, internationally. The next year's edition is usually available by September or October of the current year.

The *Air Almanac* is available as a free PDF download from the online Government Bookstore (bookstore.gpo.gov), but there are several steps involved. First you must set up an account with the bookstore, a simple process requiring only an email address and a user-selected password. Then, find the book with a search for "Air Almanac 20xx," using the year you care about. This will display several options with the same cover and title. One has "(CD-ROM)" after the title, which sells for $28. The other has nothing after the title. This is PDF version, and the one to add to your shopping cart. Then you checkout, at no charge. Once complete, click "my ACCOUNT" and you will see a link to download the book.

The *Air Almanac* for the next year is also available toward the end of the current year.

1.3 Physical and Content Descriptions

The orange hard cover copy of the almanac is 9.9" × 7" × 0.9" (25.1 cm × 17.8 cm × 2.3 cm) and weighs 1.8 lbs (0.82 kg). It is 364 pages. The Increments and Corrections pages and several other tables are printed on yellow paper at the back of the book. A loose yellow cardboard "bookmark," about equal to the page size, is included; it lists the navigational stars in both numerical order and in alphabetical order on one side, and on the other is another copy of the Altitude Corrections Table, 10° to 90°, Sun, Stars, and Planets, which is frequently referenced.

The blue paperback Commercial Edition of this book is 10" × 7" × 0.8" (25.4 cm × 17.8 cm × 2.0 cm) and weighs 1.5 lbs (0.68 k g). It is 393 pages of content with an additional 17 pages of commercial advertising at the back of the book. There is no advertising in the body of the text. A yellow cardboard "bookmark" is on a perforated page that can be removed, which is intended to match the one in the government edition.

The British Admiralty version (NP314) is a hardback cover and larger, 11.8" × 8.3" × 0.8" (30 cm × 21 cm × 2 cm), weighing 2.8 lbs (1.28 kg). It has thicker paper and larger print compared to the US versions, which could make it attractive to some navigators.

The Air Almanac is a PDF file with native page sizes of 8.5" × 11", but there are 909 pages! If this were printed two-sides on standard copy paper it would be just under 2 " thick and weigh over 3 lbs. A professional print shop could likely print it two-pages-up, reducing this by a factor of two to a book about the size the *NOAA Tide Tables: West Coast of North and South America.*

The main goal of our booklet here is to compare content and location of data in both air and nautical almanacs. This is covered, topic by topic, in the following chapters and sections. For now, however, we take a look ahead at the full table of contents of the two books on the following two pages (Sections 1.4 and 1.5) to see where we are starting. These Tables of Contents will be a convenient reference as we go through the topics.

One thing to notice when comparing these Tables of Contents, is that both books use an unusual, if not awkward, ordering of the pages. Most of this is traditional, however, so we cannot recommend any changes. Indeed, in our own textbooks we refer to *Nautical Almanac* page xxxiii for an index to the navigational stars, with the confidence that each year this information will be on this page. The *Air Almanac's* reference to the "flaps," as well as to inside and outside covers, must apply to an earlier printed version.

The *Air Almanac* PDF file has a thorough set of bookmarks, which can be opened in a left-side panel. These are useful for navigating the document, and you can add your own to these, or even rename the ones that are there. Compare those bookmarks with the Table of Contents in Section 1.4.

Next we notice that most of the topics are common to each book, though maybe described differently.

MANAC Table of Contents

1.5 NAUTICAL ALMANAC Table of Contents

INDEX TO SELECTED STARS, 2019

Name	No	Mag	SHA	Dec
			°	°
Acamar	**7**	3·2	315	S 40
Achernar	**5**	0·5	335	S 57
Acrux	**30**	1·3	173	S 63
Adhara	**19**	1·5	255	S 29
Aldebaran	**10**	0·9	291	N 17

No	Name	Mag	SHA	Dec
			°	°
1	*Alpheratz*	2·1	358	N 29
2	*Ankaa*	2·4	353	S 42
3	*Schedar*	2·2	350	N 57
4	*Diphda*	2·0	349	S 18
5	*Achernar*	0·5	335	S 57

STARS, 2019

No.	Name		Mag	SHA	Dec
				° ′	° ′
7*	*Acamar*		3·2	315 15	S 40 14
5*	*Achernar*		0·5	335 24	S 57 08
30*	*Acrux*		1·3	173 05	S 63 12
19	*Adhara*	†	1·5	255 09	S 29 00
10*	*Aldebaran*	†	0·9	290 45	N 16 33

STAR INDEX, 2019

No.	Name		Mag	SHA	Dec
				° ′	° ′
1 *	*Alpheratz*	†	2·1	357 39	N 29 12
2	*Ankaa*		2·4	353 12	S 42 12
3 *	*Schedar*		2·2	349 36	N 56 39
4 *	*Diphda*	†	2·0	348 52	S 17 53
5 *	*Achernar*		0·5	335 24	S 57 08

Figure 1-4. *Navigation Star lists from the NA (Top, pages xxxiii and bookmark) and AA (Bottom, pages F3 and F4); first 5 only, of 57 stars. The alphabetic and numeric listing are in the same table in the NA, but these are on sequential pages in the AA. The AA version has more information, in that the coordinates are given with minutes included, plus it marks with an * the stars used in Pub 249 Vol. 1, and it marks with a † the ones that can be sight reduced using Pub 249 Vols. 2 and 3—that requirement being that their declinations must be less than 29°.*

1.6 Navigational Stars

Both almanacs start out, in effect, with a list of the same 57 navigational stars, which are those stars chosen to be bright and uniformly distributed across the globe of the sky, but they present these lists in different formats. Going forward we will use the abbreviation AA for *Air Almanac*, and NA for *Nautical Almanac*.

Each of the navigational stars has a proper name and a unique number. We have frequent need to have this list sorted both numerically and alphabetically, and both almanacs include such tables, but organized differently. In the AA, on the first page of the PDF file we find the list sorted alphabetically by star name, which is repeated on page F3, whereas the list sorted by star number is postponed to page F4. In the NA, these two lists are presented side by side on the bookmark card and on page xxxiii. A section of those tables is shown in Figure 1-4.

2. GEOGRAPHICAL POSITION DATA

2.1 Overview

We come now to the key data of an almanac for celestial navigation: the locations of the celestial bodies we use in sextant sights, which are specified in terms of their Declinations (Dec) and Greenwich hour angles (GHA). These data are presented on the *daily pages*. These celestial coordinates correspond to the latitude and longitude, respectively, of the point on earth directly below the body at any moment, called the geographical position (GP). Standing at the GP, the body is directly overhead. These GPs move westward around the globe at a rate of 15° of Lon per hour, directly over a latitude equal to their declination.

The daily pages are laid out in a similar format in both books, but the big difference is the NA lists values every hour, whereas the AA lists them every ten minutes. There is no virtue to mariners in having them every 10 minutes; we still have to make minutes and seconds corrections, but this does mean that the AA daily pages section is notably bigger, especially since it only lists half a day per page compared to the NA's three pages per day. Another notable difference is the daily pages of the AA only include planets that are useful for navigation. Those too close to the sun or moon are not included. This point is discussed further in Section 4.1. A comparison of other distinctions are listed in Table 1-1, samples on the following pages.

Table 1-1. Daily Page Distinctions

Nautical Almanac	*Air Almanac*
Data every hour.	Data every 10 min.
Sun and moon on one page; stars and planets on the facing page.	All bodies on the same page.
Three days per page.	Half a day per page.
Total daily pages 243.	Total daily pages 734.
Day of the year found in Calendars Table (p 4-5).	Day of the year given on each daily page.
Navigational Star data on each daily page (GHA Aries, Dec, SHA).	GHA Aries on daily pages, Dec and SHA on pages F3 and F4.
Sun rise, set, and twilights on each daily page.	Sun rise, set, and twilights in separate tables (p A130-A145).
Planets listed every day, regardless of position relative to the sun.	Planets only listed when they are usable for sextant sights.
Minutes and seconds corrections from *Increments and Corrections Table* (p ii) for all bodies.	Minutes and seconds corrections from *several tables: sun, Aries (p A164, A165); moon, planets (p F3)*
d and v values expressed in arc minutes at the bottom of each daily page, without directions or signs.	Same units and location; called "rates," with direction given, N or S.
Moon phase as age and % illumination.	Moon phase as % illumination only, marked as waxing or waning.
Mer pass on daily pages	Mer pass in separate table

(DAY 060) GREENWICH A. M. 2019 MARCH 1 (FRIDAY) 119

UT	SUN GHA	SUN Dec.	ARIES GHA	VENUS −4.1 GHA	VENUS −4.1 Dec.	MARS 1.2 GHA	MARS 1.2 Dec.	JUPITER −2.0 GHA	JUPITER −2.0 Dec.	MOON GHA	MOON Dec.
h m	° ′	° ′	° ′	° ′	° ′	° ′	° ′	° ′	° ′	° ′	° ′
00 00	176 52.8	S 7 46.8	158 30.6	217 22	S19 30	121 19	N15 23	257 19	S22 34	239 12	S21 36
10	179 22.9	46.6	161 01.0	219 52		123 50		259 49		241 37	37
20	181 52.9	46.4	163 31.4	222 22		126 20		262 20		244 02	37
30	184 22.9	• 46.3	166 01.8	224 52	• •	128 50	• •	264 50	• •	246 27	• 37
40	186 52.9	46.1	168 32.2	227 22		131 20		267 20		248 52	37
50	189 22.9	46.0	171 02.7	229 52		133 50		269 51		251 17	37
01 00	191 53.0	S 7 45.8	173 33.1	232 22	S19 29	136 20	N15 23	272 21	S22 34	253 42	S21 37
10	194 23.0	45.7	176 03.5	234 52		138 50		274 51		256 07	37
20	196 53.0	45.5	178 33.9	237 22		141 21		277 22		258 32	37
30	199 23.0	• 45.3	181 04.3	239 51	• •	143 51	• •	279 52	• •	260 57	• 37
40	201 53.0	45.2	183 34.7	242 21		146 21		282 23		263 22	37
50	204 23.1	45.0	186 05.1	244 51		148 51		284 53		265 47	37
02 00	206 53.1	S 7 44.9	188 35.5	247 21	S19 29	151 21	N15 24	287 23	S22 34	268 12	S21 38
10	209 23.1	44.7	191 05.9	249 51		153 51		289 54		270 37	38
20	211 53.1	44.5	193 36.3	252 21		156 21		292 24		273 02	38
30	214 23.1	• 44.4	196 06.8	254 51	• •	158 51	• •	294 54	• •	275 27	• 38
40	216 53.2	44.2	198 37.2	257 21		161 22		297 25		277 52	38
50	219 23.2	44.1	201 07.6	259 51		163 52		299 55		280 17	38
03 00	221 53.2	S 7 43.9	203 38.0	262 20	S19 29	166 22	N15 24	302 25	S22 34	282 42	S21 38
10	224 23.2	43.8	206 08.4	264 50		168 52		304 56		285 07	38
20	226 53.2	43.6	208 38.8	267 20		171 22		307 26		287 32	38
30	229 23.3	• 43.4	211 09.2	269 50	• •	173 52	• •	309 56	• •	289 57	• 38
40	231 53.3	43.3	213 39.6	272 20		176 22		312 27		292 22	38
50	234 23.3	43.1	216 10.0	274 50		178 53		314 57		294 47	38
04 00	236 53.3	S 7 43.0	218 40.5	277 20	S19 28	181 23	N15 25	317 28	S22 34	297 12	S21 38
10	239 23.3	42.8	221 10.9	279 50		183 53		319 58		299 37	38
20	241 53.4	42.7	223 41.3	282 20		186 23		322 28		302 02	38
30	244 23.4	• 42.5	226 11.7	284 50	• •	188 53	• •	324 59	• •	304 27	• 38
40	246 53.4	42.3	228 42.1	287 19		191 23		327 29		306 52	38
50	249 23.4	42.2	231 12.5	289 49		193 53		329 59		309 17	38
05 00	251 53.4	S 7 42.0	233 42.9	292 19	S19 28	196 24	N15 25	332 30	S22 34	311 42	S21 38
10	254 23.5	41.9	236 13.3	294 49		198 54		335 00		314 07	38
20	256 53.5	41.7	238 43.7	297 19		201 24		337 30		316 32	38
30	259 23.5	• 41.5	241 14.1	299 49	• •	203 54	• •	340 01	• •	318 57	• 38
40	261 53.5	41.4	243 44.6	302 19		206 24		342 31		321 22	38
50	264 23.5	41.2	246 15.0	304 49		208 54		345 02		323 48	38
06 00	266 53.6	S 7 41.1	248 45.4	307 19	S19 27	211 24	N15 26	347 32	S22 34	326 13	S21 38
10	269 23.6	40.9	251 15.8	309 48		213 54		350 02		328 38	38
20	271 53.6	40.8	253 46.2	312 18		216 25		352 33		331 03	38
30	274 23.6	• 40.6	256 16.6	314 48	• •	218 55	• •	355 03	• •	333 28	• 38
40	276 53.6	40.4	258 47.0	317 18		221 25		357 33		335 53	38
50	279 23.7	40.3	261 17.4	319 48		223 55		0 04		338 18	38
07 00	281 53.7	S 7 40.1	263 47.8	322 18	S19 27	226 25	N15 26	2 34	S22 34	340 43	S21 38
10	284 23.7	40.0	266 18.3	324 48		228 55		5 04		343 08	38
20	286 53.7	39.8	268 48.7	327 18		231 25		7 35		345 33	38
30	289 23.7	• 39.6	271 19.1	329 48	• •	233 56	• •	10 05	• •	347 58	• 38
40	291 53.8	39.5	273 49.5	332 17		236 26		12 35		350 23	38
50	294 23.8	39.3	276 19.9	334 47		238 56		15 06		352 48	38
08 00	296 53.8	S 7 39.2	278 50.3	337 17	S19 26	241 26	N15 27	17 36	S22 34	355 13	S21 38
10	299 23.8	39.0	281 20.7	339 47		243 56		20 07		357 38	38
20	301 53.8	38.9	283 51.1	342 17		246 26		22 37		0 03	38
30	304 23.8	• 38.7	286 21.5	344 47	• •	248 56	• •	25 07	• •	2 28	• 38
40	306 53.9	38.5	288 52.0	347 17		251 27		27 38		4 53	38
50	309 23.9	38.4	291 22.4	349 47		253 57		30 08		7 18	38
09 00	311 53.9	S 7 38.2	293 52.8	352 17	S19 26	256 27	N15 28	32 38	S22 34	9 43	S21 38
10	314 23.9	38.1	296 23.2	354 47		258 57		35 09		12 08	38
20	316 53.9	37.9	298 53.6	357 16		261 27		37 39		14 33	38
30	319 24.0	• 37.7	301 24.0	359 46	• •	263 57	• •	40 09	• •	16 58	• 38
40	321 54.0	37.6	303 54.4	2 16		266 27		42 40		19 23	38
50	324 24.0	37.4	306 24.8	4 46		268 57		45 10		21 48	38
10 00	326 54.0	S 7 37.3	308 55.2	7 16	S19 25	271 28	N15 28	47 40	S22 34	24 13	S21 38
10	329 24.0	37.1	311 25.6	9 46		273 58		50 11		26 38	38
20	331 54.1	37.0	313 56.1	12 16		276 28		52 41		29 03	38
30	334 24.1	• 36.8	316 26.5	14 46	• •	278 58	• •	55 12	• •	31 28	• 38
40	336 54.1	36.6	318 56.9	17 16		281 28		57 42		33 54	37
50	339 24.1	36.5	321 27.3	19 45		283 58		60 12		36 19	37
11 00	341 54.1	S 7 36.3	323 57.7	22 15	S19 25	286 28	N15 29	62 43	S22 34	38 44	S21 37
10	344 24.2	36.2	326 28.1	24 45		288 59		65 13		41 09	37
20	346 54.2	36.0	328 58.5	27 15		291 29		67 43		43 34	37
30	349 24.2	• 35.8	331 28.9	29 45	• •	293 59	• •	70 14	• •	45 59	• 37
40	351 54.2	35.7	333 59.3	32 15		296 29		72 44		48 24	37
50	354 24.2	35.5	336 29.8	34 45		298 59		75 14		50 49	37
Rate	15 00.1	N0 00.9		14 59.4	N0 00.5	15 00.8	N0 00.6	15 02.2	0 00.0	14 30.2	0 00.0

Lat.	Moonrise	Diff.
N		
°	h m	m
72	■	*
70	■	*
68	07 41	+24
66	06 30	+24
64	05 54	+25
62	05 28	+25
60	05 08	+25
58	04 52	+25
56	04 37	+25
54	04 25	+25
52	04 14	+25
50	04 05	+25
45	03 45	+25
40	03 28	+25
35	03 15	+25
30	03 03	+25
20	02 42	+25
10	02 24	+25
0	02 08	+25
10	01 51	+25
20	01 34	+25
30	01 13	+24
35	01 01	+25
40	00 48	+24
45	00 32	+24
50	00 12	+24
52	00 02	+24
54	24 43	+27
56	24 31	+27
58	24 17	+28
60	24 01	+28
S		

Moon's P. in A.

Alt	Corr	Alt	Corr
°	+ ′	°	+ ′
0		56	
	55		30
10		57	
	54		29
15		59	
	53		28
18		60	
	52		27
21		61	
	51		26
24		62	
	50		25
26		63	
	49		24
29		64	
	48		23
31		66	
	47		22
33		67	
	46		21
34		68	
	45		20
36		69	
	44		19
38		70	
	43		18
40		71	
	42		17
41		72	
	41		16
43		73	
	40		15
44		74	
	39		14
46		75	
	38		13
47		76	
	37		12
48		78	
	36		11
50		79	
	35		10
51		80	
	34		
52			
	33		
54			
	32		
55			
	31		
56			
	30		
57			

Sun SD 16′.1
Moon SD 15′
Moon ill. 24% –

Figure 2-1. *An* Air Almanac *daily page. The following page has times 1200 to 2350.*

The examples provided below show that although the tables are laid out differently and the corrections are organized differently we get the same Dec and GHA data from both, and it would be difficult to say which is easier or more intuitive.

2.1 Finding Declination: Sun, Moon, Planet, Star

For samples we use a time of 06h 44m 45s UTC on March 1, 2019.

Sun Declination

From the AA daily page 119 (Figure 2-1) we find at 06h 40m the sun's Dec is S 7° 40.4'. At the bottom of the Dec column we see it has a rate of change of "N0 00.9," which means it changes toward the north at a rate of 0° 0.9' per hour. This is a South declination changing to the north, so the value muse be getting smaller, which in turn implies any correction we apply based on this rate will be negative.

Here, however, we can fall back onto standard procedures used in the NA to check this, because the direction of this rate is not given in the NA. Namely, we look at the next tabulated Dec value (06h 50m in this case), which is N 7° 40.3', so the declination is indeed decreasing with time, and we would therefore label this rate as negative, meaning the correction is going to negative. (In NA terms, we have d = -0.9.)

Our desired time is 06h 44m 45s; we know the value at 06h 40m; now we have to find the correction, meaning here how much does this declination decrease in 4m 45s if it is decreasing at a rate of 0.9' per hour.

Here we come across a minor difference between AA and NA. The NA has a table for this correction, the AA does not. The AA simply ignores this correction, even though they do give us the rate needed to apply it. The idea is simply that given the values every 10 min, this correction will be small compared to what might occur in the NA where the values are only every hour. That is indeed true, and the AA practice of ignoring this correction is perfectly valid since we get the values every 10 minutes. The example here is near an equinox when the rate is near 1.0' per hour, which is the largest it ever gets. (Near the equinoxes, the GP of the sun is moving toward or away from the equator at a speed of 1 kt.) Most of the year this rate is much lower and this is even less of an issue.

To verify this, the largest increment we can get is 5 min, halfway between two tabulated values. The largest the rate will be 1.0' per hour. So the largest this correction can be is 5 min × 1'/60 min = 0.08'. This is within the accuracy of the almanac data itself, so we can ignore this.

The AA value we get in this one step (S 7° 40.4') is the same that will be found from the NA, which lists the 06h value as S 7° 41.1' with a d-value (rate) of -0.9'. From the NA Increments and Corrections Table for 44m 45s we find a correction of -0.7', yielding the same value S 7° 40.4'.

In short, the AA practice of tabulating values every 10 min, removes the need for an increments correction on the sun's declination.

Table 2-2. Compare AA and NA Declinations, 06:44:45 Mar 1, 2019				
Almanac	*Air*		*Nautical*	*Difference*
Body	*Declination*	*Rate (d-value)*	*Declination*	
Sun	S 7° 40.4'	-0.9	S 7° 40.4'	0.0
Moon	S 21° 38'	0.0	S 21° 38.5'	0.5'
Venus	S 19° 27'	+0.5	S 19° 27.4'	0.4'
Mars	N 15° 26'	+0.6	N 15° 25.7'	0.3'
Jupiter	S 22° 34'	0.0	S 22° 33.9'	0.1'
Achernar (#5) See Notes	(a) S 57° 08' (b) S 57° 08.7'	—	S 57° 08.8'	0.8' 0.1'
Aldebaran (#10) See Notes	(a) N 16° 33' (b) N 16° 32.7'	—	N 16° 32.7	0.3' 0.0

Table Notes: *(a) From AA navigational star lists (page F3, F4); (b) From AA extra star list (page A158) which has monthly values to the tenth of an arc minute. In this comparison, the Nautical Almanac values are more accurate, as they are based on data with more precision.*

Moon and Planets Declination

The AA treats the declination of moon and planets just as it does the sun, namely, providing the data every 10 minutes, they do not include any increments corrections. We can test this with the same time and date as above. On March 1, 2019 at 06h 44m 45s, we start with the values at 06h 40m, which the AA assumes is equivalent to the values at 06h 44m 45s. Declinations are compared in Table 2.2. The AA Explanation Section 25 (page A17) states: "No interpolation is required for Dec.; the maximum error cannot exceed 2′.0 for the Moon, or 0′.8 for the Sun, planets and stars." Section 4.1 explains the AA daily pages planet display conventions.

Star Declinations

Star declinations do not vary on a daily basis, but do vary cyclically by small amounts throughout the year, although at different rates for different stars. Referring to Table 2-2, Achernar's declination minutes vary throughout 2019 from 8.1' to 8.9'; Aldebaran's varies from 32.6' to 32.8'. The NA provides star declinations daily to cover that. Declination minutes are not listed in the AA daily pages nor on the navigational star lists, but the AA includes the same list of 173 Selected Stars (page A158-A163) that is included in the NA (page 268-273). This list includes the numbered navigational stars, and this list has the declination minutes to the tenth throughout the year. Thus we can find more accurate star declinations from that table.

The selected star tables are ordered by star SHA, with half the year using a Greek-letter nomenclature for the stars, and the other half of the year using their proper names—we need to stare at it a bit to sort that out! The numbers of the navigational stars are included. Again, this table is identical in AA and NA.

2.2 Finding GHA: Sun, Moon, Planet, Star

We find these values to the nearest 10 minutes on the daily pages (Figure 2-1) and then make a correction for the remaining minutes and seconds. Again, we use an example time of 06h 44m 45s UTC on March 1, 2019.

Sun GHA and Aries

Note from the daily pages (Figure 2-1) that the sun and Aries data are given to the nearest 0.1' of angle, whereas moon and planets are given only to nearest 1'. This implies that the final accuracy of sun and Aries data will be higher, but this is rarely a practical limitation to the ultimate celestial fix accuracy.

From the AA daily pages for 06h 40m we find GHA of 276° 53.6', which leaves the correction for 4m 45s for a GP that is moving west at a rate of 15° 0.1' per hour, listed at the bottom of the GHA column. We find these corrections in the Interpolation of GHA Sun Table (page A164), shown in Figure 2-2. The correction found there is 1° 11.3', so the final GHA of the sun is 276° 53.6 + 1° 11.3' = 278° 4.9'.

In the NA we look this up for 06h (266° 53.6'); then add the correction (11° 11.3') for 44m 45s from the Increments and Corrections Table, for a final value of 278° 4.9', which agrees.

From the AA daily pages, GHA of Aries at 06h 40m is 258° 47.0'. The correction for 4m 45s is taken from the GHA Interpolation of GHA Aries Table (page A165, which is similar to the one for the sun shown in Figure 2-2 but with slightly different values). The correction is +1° 11.4'. The final value is then 258° 47.0 + 1° 11.4' = 259° 58.4'.

A164 INTERPOLATION OF G.H.A. SUN

s	0m	1m	2m	3m	4m	5m	6m	7m	8m	9m	s
	° ′	° ′	° ′	° ′	° ′	° ′	° ′	° ′	° ′	° ′	
00	0 00.0	0 15.0	0 30.0	0 45.0	1 00.0	1 15.0	1 30.0	1 45.0	2 00.0	2 15.0	00
01	0 00.3	0 15.3	0 30.3	0 45.3	1 00.3	1 15.3	1 30.3	1 45.3	2 00.3	2 15.3	01
02	0 00.5	0 15.5	0 30.5	0 45.5	1 00.5	1 15.5	1 30.5	1 45.5	2 00.5	2 15.5	02
03	0 00.8	0 15.8	0 30.8	0 45.8	1 00.8	1 15.8	1 30.8	1 45.8	2 00.8	2 15.8	03
45	0 11.3	0 26.3	0 41.3	0 56.3	1 11.3	1 26.3	1 41.3	1 56.3	2 11.3	2 26.3	45
46	0 11.5	0 26.5	0 41.5	0 56.5	1 11.5	1 26.5	1 41.5	1 56.5	2 11.5	2 26.5	46
47	0 11.8	0 26.8	0 41.8	0 56.8	1 11.8	1 26.8	1 41.8	1 56.8	2 11.8	2 26.8	47
48	0 12.0	0 27.0	0 42.0	0 57.0	1 12.0	1 27.0	1 42.0	1 57.0	2 12.0	2 27.0	48
49	0 12.3	0 27.3	0 42.3	0 57.3	1 12.3	1 27.3	1 42.3	1 57.3	2 12.3	2 27.3	49
50	0 12.5	0 27.5	0 42.5	0 57.5	1 12.5	1 27.5	1 42.5	1 57.5	2 12.5	2 27.5	50
51	0 12.8	0 27.8	0 42.8	0 57.8	1 12.8	1 27.8	1 42.8	1 57.8	2 12.8	2 27.8	51
52	0 13.0	0 28.0	0 43.0	0 58.0	1 13.0	1 28.0	1 43.0	1 58.0	2 13.0	2 28.0	52
53	0 13.3	0 28.3	0 43.3	0 58.3	1 13.3	1 28.3	1 43.3	1 58.3	2 13.3	2 28.3	53
54	0 13.5	0 28.5	0 43.5	0 58.5	1 13.5	1 28.5	1 43.5	1 58.5	2 13.5	2 28.5	54
55	0 13.8	0 28.8	0 43.8	0 58.8	1 13.8	1 28.8	1 43.8	1 58.8	2 13.8	2 28.8	55
56	0 14.0	0 29.0	0 44.0	0 59.0	1 14.0	1 29.0	1 44.0	1 59.0	2 14.0	2 29.0	56
57	0 14.3	0 29.3	0 44.3	0 59.3	1 14.3	1 29.3	1 44.3	1 59.3	2 14.3	2 29.3	57
58	0 14.5	0 29.5	0 44.5	0 59.5	1 14.5	1 29.5	1 44.5	1 59.5	2 14.5	2 29.5	58
59	0 14.8	0 29.8	0 44.8	0 59.8	1 14.8	1 29.8	1 44.8	1 59.8	2 14.8	2 29.8	59
60	0 15.0	0 30.0	0 45.0	1 00.0	1 15.0	1 30.0	1 45.0	2 00.0	2 15.0	2 30.0	60

Figure 2-2. *Interpolation table for sun's GHA. For 4m 45s, the correction is + 1° 11.3'.*

INTERPOLATION OF GHA F3

Increment to be added for intervals of UT to GHA of Sun, Aries and planets; Moon.

SUN, etc. m s	° ′	MOON m s	SUN, etc. m s	° ′	MOON m s	SUN, etc. m s	° ′	MOON m s
00 00		00 00	03 17		03 25	06 37		06 52
	0 00			0 50			1 40	
01		00 02	21		03 29	41		06 56
	0 01			0 51			1 41	
05		00 06	25		03 33	45		07 00
	0 02			0 52			1 42	
09		00 10	29		03 37	49		07 04
	0 03			0 53			1 43	
13		00 14	33		03 41	53		07 08
	0 04			0 54			1 44	
17		00 18	37		03 45	06 57		07 13
	0 05			0 55			1 45	
21		00 22	41		03 49	07 01		07 17
	0 06			0 56			1 46	
25		00 26	45		03 54	05		07 21
	0 07			0 57			1 47	
29		00 31	49		03 58	09		07 25
	0 08			0 58			1 48	
33		00 35	53		04 02	13		07 29
	0 09			0 59			1 49	
37		00 39	03 57		04 06	17		07 33
	0 10			1 00			1 50	
41		00 43	04 01		04 10	21		07 37
	0 11			1 01			1 51	
45		00 47	05		04 14	25		07 42
	0 12			1 02			1 52	
49		00 51	09		04 19	29		07 46
	0 13			1 03			1 53	
53		00 55	13		04 23	33		07 50
	0 14			1 04			1 54	
00 57		01 00	17		04 27	37		07 54
	0 15			1 05			1 55	
01 01		01 04	21		04 31	41		07 58
	0 16			1 06			1 56	
05		01 08	25		04 35	45		08 02
	0 17			1 07			1 57	
09		01 12	29		04 39	49		08 06
	0 18			1 08			1 58	
13		01 16	33		04 43	53		08 11
	0 19			1 09			1 59	
17		01 20	37		04 48	07 57		08 15
	0 20			1 10			2 00	
21		01 24	41		04 52	08 01		08 19
	0 21			1 11			2 01	
25		01 29	45		04 56	05		08 23
	0 22			1 12			2 02	
29		01 33	49		05 00	09		08 27
	0 23			1 13			2 03	
33		01 37	53		05 04	13		08 31
	0 24			1 14			2 04	
37		01 41	04 57		05 08	17		08 35
	0 25			1 15			2 05	
41		01 45	05 01		05 12	21		08 40
	0 26			1 16			2 06	
45		01 49	05		05 17	25		08 44

Figure 2-3. *Interpolation table for sun, Aries, and planets on the left (SUN, etc.), and Moon on the right. Use the tabulated value that is within the minutes and seconds interval of the observation time.*

For a sight at 06h 44m 45s, the daily page time is 06h 40m, and we enter here with 4m 45s. For the sun, Aries, or a planet, this correction would be 1° 11′—we might be tempted to average the upper and lower values to get 11.5′, especially since we know from the sun interpolation table that the answer is 11.3′, but that would be wrong. The Explanation states that when the time interval matches a tabulated value, we should take the upper value; in this case 1° 11′. This would apply mostly to moon, and planets, as it is more logical to use the interpolation tables on pages A164 and A165 for the sun and Aries.

In short, this table from page F3 is effectively a GHA interpolation table for planets (left column) and moon (right column).

In the NA at 6h we find GHA Aries = 248° 45.4' and in the Increments and Corrections Table for 44m 45s we get 11° 13.1' for a total GHA of 259° 58.5'. This comparison and other values are summarized in Table 2-3.

Moon and Planet GHA

The 6h 40m value of the GHA of the moon from the AA daily pages (Figure 2-1) is 335° 53', given rounded to the nearest whole minute, unlike the sun and Aries values. The correction for 4m 45s we find in the GHA Interpolation Tables (page F3; Figure 2-3), which has one column for sun, Aries, and planets and another column for the moon. The sun and Aries corrections from this table, however, are redundant with the tables we used earlier for sun (page A164) and Aries (page A165), and indeed this one is less precise and not quite as easy to use. Consequently, as noted in Figure 2-3, this is effectively a GHA correction table for just moon (right column) and planets (left column).

Referring to Figure 2-3, the 4m 45s correction for the moon's GHA is +1° 09', so the final value is 335° 53' + 1° 09 = 337° 02'.

Table 2-3. Compare AA and NA GHA, 06:44:45 Mar 1, 2019			
Almanac	*Air*	*Nautical*	*Difference*
Body	*GHA*	*GHA*	
Sun	278° 4.9'	278° 4.9'	0.0
Aries	259° 58.4'	259° 58.5'	0.1'
Moon	337° 02'	337° 1.5'	0.5'
Venus	318° 29'	318° 29.5'	0.5'
Mars	222° 36'	222° 36.2'	0.2'
Jupiter	358° 44'	358° 44.8'	0.8'
Acamar (#7) See Notes	(a) 215° 13.4' (b) 215° 14.0'	215° 14.0'	0.6' 0.0'
Aldebaran (#10) See Notes	(a) 190° 43.4' (b) 190° 43.4'	190° 43.4'	0.0' 0.0

Table Notes: *(a) From AA navigational star lists (page F3, F4); (b) From AA extra star list (page A158) which has monthly values to the tenth of an arc minute. In this comparison, the Nautical Almanac values are more accurate, as they are based on data with more precision.*

In the NA we find at 06h a GHA of 326° 12.5' with v = 11.2', and then in the Increments and Corrections Table, using 44m 45s, we find a correction of 10° 40.7' plus a v correction of 8.3' for a total value of the GHA = 337° 1.5'. These data are added to Table 2-3 on comparing GHA values.

The planet GHA values for Venus, Mars, and Jupiter from the AA daily pages (6h 40m) are, respectively, 317° 18', 221° 25', 357° 33', again precise just to the whole minute. The 4m 45s corrections to these from page F3 ("SUN, etc." column) is +1° 11' yielding the final values shown in Table 2-3.

GHA of Stars

The GHA of a star is equal to the GHA of Aries plus the sidereal hour angle (SHA) of the star. GHA Aries is on the AA daily pages, which agree with the NA values, so the question is where to best find the SHA of a particular star. We are back the same situation as when finding declination. We can get the SHA in the navigational star lists (pages F3 and F4), which is a yearly average, or we can get these values monthly from the Selected Stars list on page A158.

Using the daily page value of GHA Aries (06h 40m), corrected for 4m 45s using the Interpolation of GHA Aries Table (Page A165), we get the value shown in Table 2-3. The respective SHA values for two sample stars from the navigational star lists (pages F3 and F4; Figure 1-4) are: Acamar 315° 15' and Aldebaran 290° 45'. The resulting GHAs are Acamar 259° 58.4' + 315° 15 = 575° 13.4' = 215° 13.4'. For Aldebaran it is 259° 58.4' + 290° 45' = 550° 43.4' = 190° 43.4'. These AA values are labeled (a) in Table 2-3.

Looking up values in the NA, we see the GHA Aries is 0.1' larger; SHA for Acamar is 0.5' larger; SHA for Aldebaran is 0.1' smaller, which yields the NA values shown in Table 2-3.

As noted earlier, we can get more accurate values from the AA by using the Selected Stars Table for March, which gives Aldebaran minutes 45.0' and for Acamar 15.6'. We see in the values marked (b) in Table 2-3 that it is better for both declination and for SHA to get AA star data from the Selected Stars list, rather than from the navigational stars lists. Here we present just this one test on one day, but the results are consistent with the accuracy summary in the AA Explanation section.

The primary conclusion here is the basic astro data needed for cel nav, namely GHA and declination, can be obtained from the AA just as well as the NA. The accuracy may on the average be a bit lower for the moon and planets, but this difference would rarely interfere with the final accuracy of the celestial fix which is dependent on other factors, not the least of which is the correction for the motion of the vessel during the sight sessions. We also see that the process of extracting the data from the AA is simpler is several cases where extra corrections are not needed because the data are presented every 10 min instead of every hour as in the NA.

3. LOCAL MEAN TIME OF EVENTS

The presentation of event times is where the two almanacs differ the most. We can get the same results, but the tables used and correction procedures are different.

3.1 Sunrise, Sunset, and Twilights

In the AA, sunrise, sunset, and twilight times are listed as a function of date and latitude in a sequence of tables spanning pages A130 to A145. Like the NA, the AA data are only given every 3 days, and the latitudes listed are identical in both almanacs. The days listed are the same as well, but this is less obvious.

In the NA, the rise and set data are on the daily pages. There are three days per page, but only one set of rise and set data. That data applies to the middle day. Comparing middle days of the NA with dates listed in the AA, we see they are the same. This means that in both almanacs, if the date you want is not listed in the AA, meaning it will not be a middle day of the NA, then we must extract two values (day before and day after) and average them. We leave that as an exercise, and for our example use a date that is listed, namely July 10, 2019, and we will consider our present latitude as 30° N. The data are the same in both almanacs, so the results should be the same.

Conversion of Arc to Time

For July 10, at 30N, sunrise is 0506, and sunset is 1904. These are *local mean times* (LMT), which means they are the UTC of the event as observed on the Greenwich meridian. The UTC of the event observed at other longitudes will be this time plus the observer's Lon West (converted to time), or minus Lon East (converted to time). The Conversion of Arc to Time tables are the same in both almanacs, except that the NA version includes decimal parts of 1' (0.25', 0.50', and 0.75'), but the AA version does not. These are rarely called for, so this is not a significant difference. In short, sunrise and sunset is the same in both almanacs.

Twilight Times

The conventions on twilight time presentation are not the same in the two almanacs, but first a reminder of the definitions. *Civil twilight* is the specific time of day when the sun is 6° below the horizon (zenith distance = 96°), typically indicating when it is just dark enough to see the brightest stars in the evening, or too bright to see them in the morning. *Nautical twilight* is the specific time of day when the sun is 12° below the horizon (zenith distance = 102°), typically indicating it is too dark to see the horizon. Losing sight of the horizon ends the evening sight session; first seeing the horizon starts the sight session in the morning. *Evening twilight* is a time period from civil twilight (dark enough to see bright stars) to nautical twilight (too dark to see the horizon). *Morning twilight* is likewise a time period from nautical twilight (horizon first visible) to civil twilight (bright stars fade).

The NA lists the times of civil and nautical twilight in both the morning and the evening in a logical sequential order, meaning in the morning: nautical twilight, civil twilight, and sunrise, followed in the evening with: sunset, civil twilight, and nautical twilight. The AA does not do it that way.

The AA has a sunrise table, followed by a sunset table, followed by morning civil twilight, followed by evening civil twilight. Morning civil twilight is the end of the morning twilight period of sight taking, and evening civil twilight is the beginning of the evening twilight period of sight taking. The logic of this presentation must be related to air navigation. The two civil twilight times given do indeed match the NA values, but we do not have from this a direct estimate of the length of the sight taking session... as experienced from the surface. There are numerous special tables in the AA for illumination periods at various altitudes, but that does not help us on the surface.

Without an NA, we can compute the LMT of nautical twilight by finding the LHA of the sun that corresponds to a zenith distance (z) of 102°. Then divide that angle by 15 to covert it to hours, and add that to the meridian passage time. The formula is:

LMT of twilight = MER PASS (hr) +
(1/15)*ACOS[(COS(z)-SIN(Lat)*SIN(Dec))/(COS(Lat)*COS(Dec))],

where z = 96 for civil twilight and z = 102 for nautical twilight. Here is a test case: Dec = 22.88°, Lat = 50°; z = 102, LMT = 21h 58.6m; z = 96, LMT = 20h 55.2m.

Or we can put this in perspective and move on. That is: even if we do not know how long the working twilight lasts on our first night of sextant sights, we will learn it then, as we almost always use up the full period for sights. Once back to the nav station, we can just look at the time interval between first and last sights, and that will indeed be the time we have, essentially for the full crossing. Although it will get shorter headed toward the equator.

3.2 Local Apparent Noon (Meridian Passage) A151

The times of meridian passage in the AA are given in Table 1, Meridian Passage and Declination of the Sun at 12h UTC, on pages A151 and A152. The times are given relative to 1200. For example, on July 10 the listing is +5, meaning the time of meridian passage is 1205. On Sept 26, it is -9, meaning LAN occurs at 1151.

This is an especially convenient table in that it lists the declination of the sun as well, and the main reason we want the time of LAN is to prepare for a noon sight, and to analyze that sight, once done, all we need is the declination of the sun, which is right here.

In the NA, "mer pass" time is given on each daily page, specified for each of the three days on the page. It is in a small box on the bottom right of the sun-moon daily page. It is given at both 00h and 12h, but that is more confusing than helpful. The 00h value might be a historic artifact related to tide prediction. I am not sure.

The NA also lists the *equation of time* in this same box, which can be used to get the mer pass time accurate to the second. On July 10, 2019, for example, the NA shows mer pass time at 1205 and the equation of time as 5m 22s. Thus the more precise time of LAN would be 12h 05m 22s. This is not often called for in routine celestial navigation, but it could have some application in shortcut or emergency methods.

Generally any special need of this equation of time can be solved with the GHA of the sun at 12h, which is in the AA. On July 10 at 12h it is 358° 39.4', which means the sun has not yet reached the meridian at 1200. It has to travel 360° - 358° 39.4 = 1° 20.6' = 80.6' farther, which we can convert to time at 15'/1 min or 80.6/15 = 5.37 min = 5m 22.4s.

3.3 Moonrise and Moonset

All required moon forecasting is indeed in the AA and it leads to the same results as found from an NA, but from a practical perspective, this is rarely used for actual cel nav related thinking. The visibility of the moon is more a question of night watch planning or for planning a nighttime landfall. A bright moon is always helpful—and, indeed, the more help it is in seeing, the less attractive it is for a celestial sight target, because that extra light distorts the horizon below it, bringing it too close to the navigator.

On the other hand, the bright moon interference issue is more a factor when the moon is less that halfway up the sky. A high bright moon can on occasion be a bonus in that it illuminates enough of the horizon around you that you can take nighttime star sights. Any single one will not be as good as we would get with a good twilight horizon, but with three well spaced ones with equally bad horizons below them, we can still get a good fix.

With that said, moon rise and set data are on the AA daily pages, with rise times on the AM pages, and set times on the PM pages. On July 10, 2019 at 30N, moonrise is at 1312; the same value found in the NA.

This is a local mean time, but these moon predictions are complicated by the motion of the moon around the earth as the earth rotates daily on its axis. Unlike similar times for the sun, the interval between successive moonrises or moonsets varies (being generally greater than 24h), so times of rise or set must be interpolated from their tabular values on the Greenwich meridian to the meridian of the observer. This is facilitated in the AA with a parameter called "Diff." It represents one half of the difference between rise or set time on a particular day, and that of the following day in west Lon, or the previous day in east Lon.

For example, let's consider a position of 30N, 140W on July 10, 2019. The rise time on the 10th is 1312 and the rise time on the 11th is 1413, which is a difference of 61 min, and half of that is 30.5 min. In the AA we find a Diff. value of 31 next to the rise time, which we use with our Lon of 140W to enter the Interpolation of Moonrise, Moonset for Longitude Table on page F4 to find a correction of +23 min. Therefore the correct LMT of moonrise on July 10th at 140W is 1312 + 0023 = 1335. At 140 W, the UTC of the event would be 1335 + 0920 (140° converted to time) = 2255 UTC.

In the NA, we do something very similar but we have to compute the Difference in the two rise times ourselves, it is not tabulated as in the AA, and the NA table for the interpolation (Table II on page xxxii) uses this full difference of 61 min, not the half difference used in the AA.

Now with that nasty business covered, we should note that chances are you will never ever have to make that type of correction... unless you are a candidate for a USCG unlimited masters license, and you are unlucky enough to get that question on your exam. And for that, you will have to use the NA, even though the AA process is a step or two shorter, as it tabulates the Diff. factor.

4. STAR AND PLANET IDENTIFICATION

4.1 Only Useful Planets on the Daily Pages

One significant difference between AA and NA that could be either noticed immediately, or not noticed at all, depending on what daily page you look at, is that the AA only includes those planets that are useful for sextant sights. There are four visible planets in this category, but on some days they are too close to the sun or moon to be useful. When that happens, the AA simply does not list them or it leaves a note in the planet column saying why it is not there, as shown in Figure 4-1. Those used to the NA expect to see almanac data for all planets, regardless of whether they are useful or not.

4.2 Star Charts

Each of the almanacs, AA and NA, have unique star charts. There is one in the AA and two in the NA, but none is redundant; they all complement each other. The star chart from the AA is actually the more famous with navigators, because it focuses on the navigational stars and is arguably the most useful. It is distributed by the USNO separately, and many users of the NA have a copy, although few may

734 (DAY 367) GREENWICH P. M. 2020 JANUARY 2 (THURSDAY)

UT	SUN		ARIES	VENUS −4.0		MARS 1.6				MOON		Lat.	Moon-set	Diff.
	GHA	Dec.	GHA	GHA	Dec.	GHA	Dec.	GHA	Dec.	GHA	Dec.			
h m	° ′	° ′	° ′	° ′	° ′	° ′	° ′	° ′	° ′	° ′	° ′			
12 00	359 03.1	S22 56.1	281 35.8	322 19	S17 42	44 20	S19 41			275 53	S 3 15	N		
10	1 33.0	56.1	284 06.2	324 49		46 50				278 18	13	°	h m	m
20	4 03.0	56.0	286 36.6	327 19		49 20		THE		280 44	11	72	23 42	+53
30	6 32.9	• 56.0	289 07.0	329 49	• •	51 50	• •			283 10	• 09	70	23 44	+49
40	9 02.9	56.0	291 37.4	332 19		54 20		REMAINING		285 36	07	68	23 45	+46
50	11 32.8	55.9	294 07.8	334 49		56 50				288 02	05	66	23 46	+43
13 00	14 02.8	S22 55.9	296 38.2	337 19	S17 41	59 20	S19 41	NAVIGATIONAL		290 28	S 3 03	64	23 47	+41
10	16 32.7	55.9	299 08.6	339 49		61 51				292 54	02	62	23 48	+39
20	19 02.7	55.8	301 39.0	342 18		64 21		PLANETS		295 20	3 00	60	23 48	+38
30	21 32.6	• 55.8	304 09.5	344 48	• •	66 51	• •			297 45	2 58	58	23 49	+37
40	24 02.6	55.8	306 39.9	347 18		69 21		ARE		300 11	56	56	23 50	+36
50	26 32.6	55.7	309 10.3	349 48		71 51				302 37	54	54	23 50	+34
14 00	29 02.5	S22 55.7	311 40.7	352 18	S17 40	74 21	S19 42	TOO		305 03	S 2 52	52	23 50	+33
10	31 32.5	55.6	314 11.1	354 48		76 51				307 29	50	50	23 51	+33
20	34 02.4	55.6	316 41.5	357 18		79 21		CLOSE		309 55	48	45	23 52	+30
30	36 32.4	• 55.6	319 11.9	359 48	• •	81 51	• •			312 21	• 46	40	23 52	+29
40	39 02.3	55.5	321 42.3	2 18		84 22		TO		314 47	44	35	23 53	+28
50	41 32.3	55.5	324 12.7	4 48		86 52				317 12	42	30	23 53	+26
15 00	44 02.2	S22 55.5	326 43.2	7 17	S17 39	89 22	S19 42	THE		319 38	S 2 40	20	23 54	+24
10	46 32.2	55.4	329 13.6	9 47		91 52				322 04	38	10	23 55	+22
20	49 02.1	55.4	331 44.0	12 17		94 22		SUN		324 30	36	0	23 56	+21
30	51 32.1	• 55.3	334 14.4	14 47	• •	96 52	• •			326 56	• 35	10	23 57	+19
40	54 02.0	55.3	336 44.8	17 17		99 22		FOR		329 22	33	20	23 57	+17
50	56 32.0	55.3	339 15.2	19 47		101 52				331 48	31	30	23 58	+15
16 00	59 01.9	S22 55.2	341 45.6	22 17	S17 38	104 23	S19 43	OBSERVATION		334 14	S 2 29	35	23 58	+14
10	61 31.9	55.2	344 16.0	24 47		106 53				336 39	27	40	23 59	+12
20	64 01.8	55.2	346 46.4	27 17		109 23				339 05	25	[illegible]	[illegible]	[illegible]

Figure 4-1. *AA daily page showing Saturn and Jupiter missing because they are not useful for sextant sights at this time because they are "too close to the sun" for useful sextant sights. In physical space, they are not at all close to the sun; this just means they are in the same direction as the sun. This can apply to any of the planets. This is a nice tip for navigators that we do not get in the NA. Note this is Day 367. To fill out the daily pages table format they include an extra day or two. This year in the NA, it went only to Jan 1, day 366.*

know of its origin in the AA. The AA Explanation section has an in-depth discussion of the use of this map. The AA sample is shown in Figure 4-2. The northern sky version of the NA star map is shown in Figure 4-3.

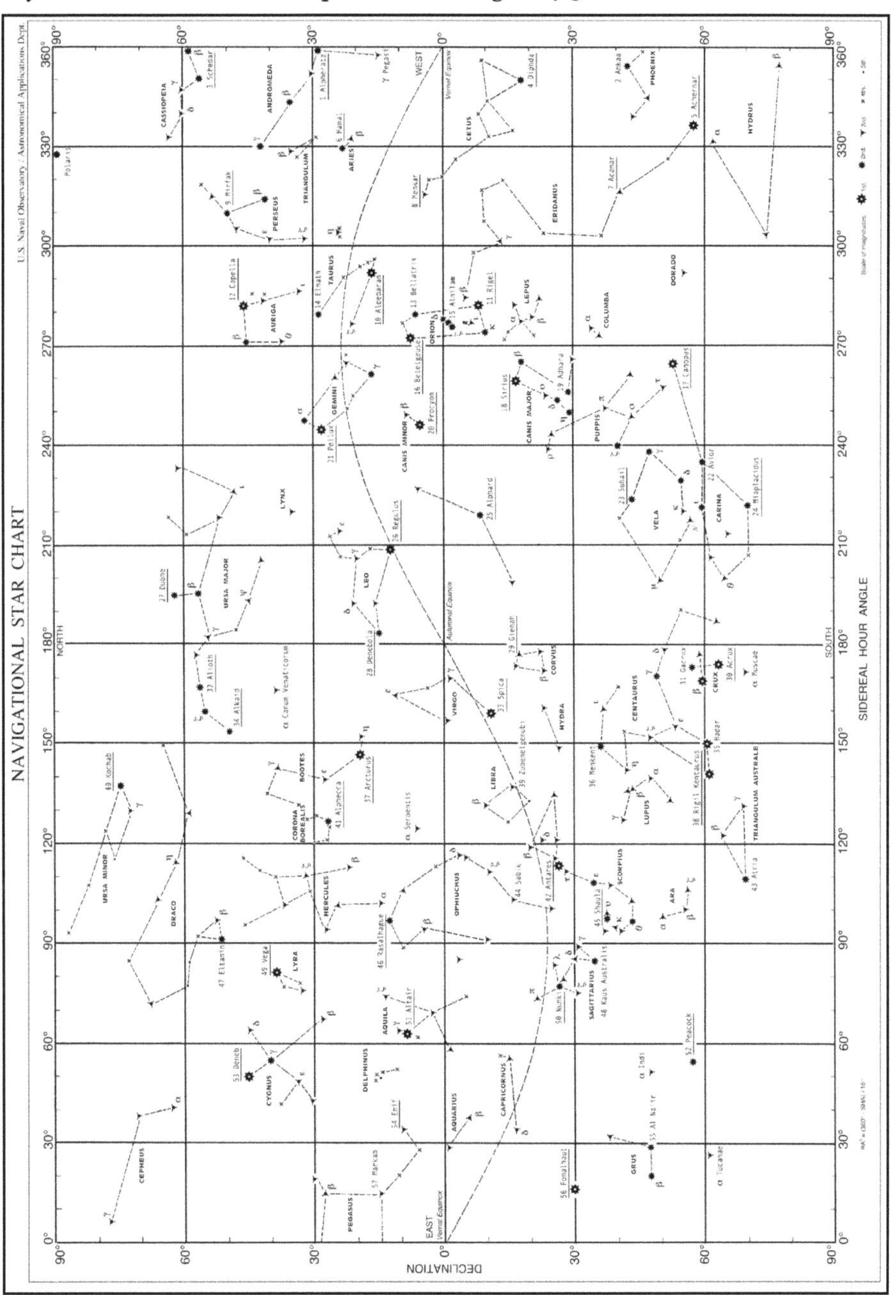

Figure 4-2. *AA Star Chart showing positions of the navigational stars (numbered and underlined). To find the point overhead, draw a horizontal line at a declination equal to your latitude and a vertical line as SHA = Lon W – GHA Aries, or SHA = (360 – Lon E) – GHA Aries. (If SHA is negative, add 360°.) See related discussion in Appendix A2.*

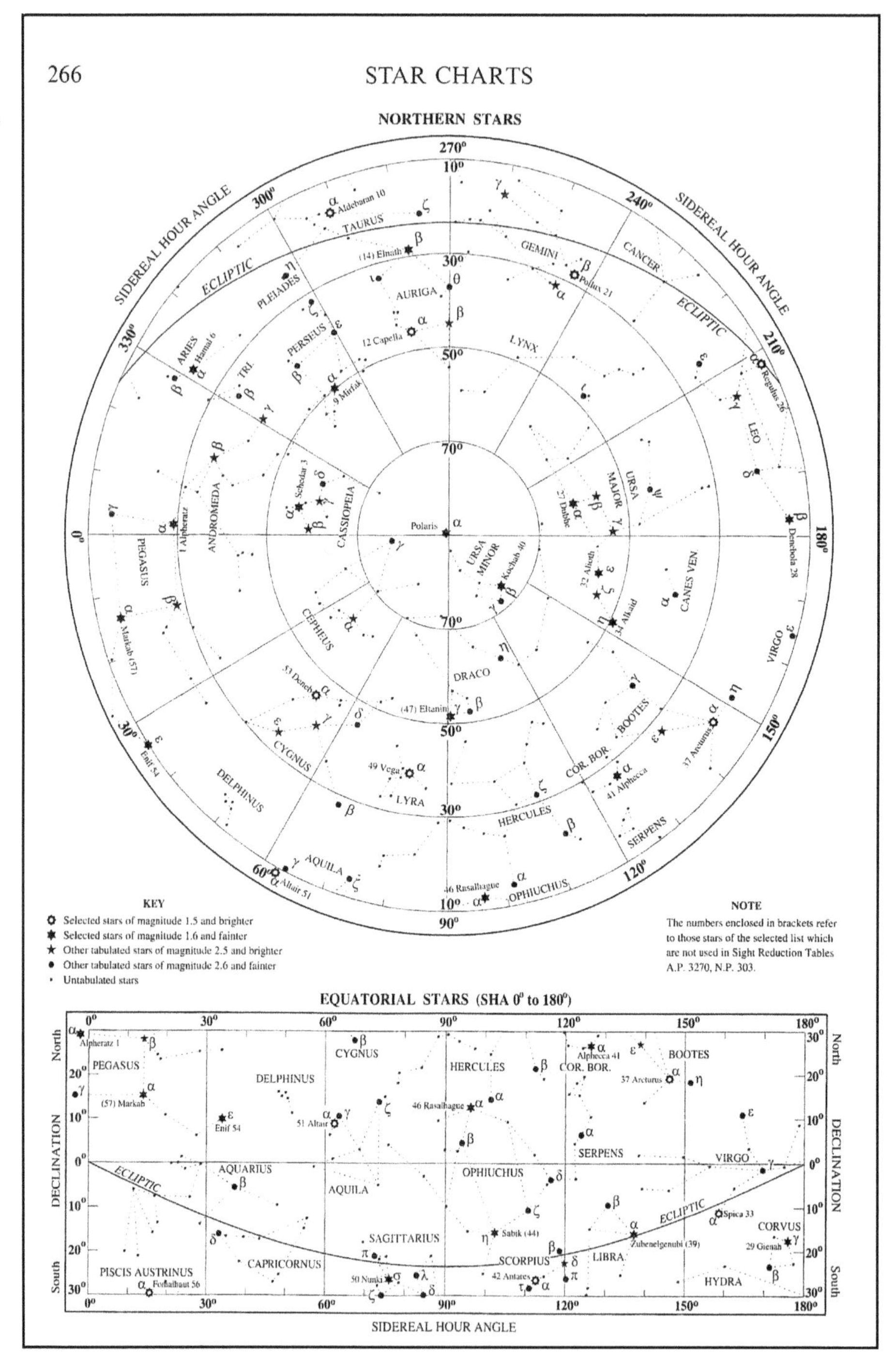

Figure 4-3. *Star Charts from the NA. This is for the northern sky and equatorial region below it. There are another two for the southern sky. The navigational stars are numbered.*

4.3 Planet Diagrams, Notes, and Do Not Confuse

Both almanacs have extended sections on planet identification, including diagrams designed to help identify planets. The diagrams are similar in appearance, but designed differently in several ways. The AA version takes up two pages (A122 and A123) and has more information included (Figure 4-5), such as moon phases and relative positions. The NA version (page 9) is shown in Figure 4-6, which includes a twilight band not shown on the AA version. Planet diagrams are a quick way to note when a planet will be an evening or morning star.

A table or notes section telling which stars and planets might be confused is also common to both almanacs; the NA cautions about planet confusion; the AA cautions about star-planet confusion, but in a different format as shown in Figure 4-4.

Both books have useful information on visibility of stars and planets throughout the almanac year, and a navigator would benefit from reading both treatments.

DO NOT CONFUSE

Venus with Jupiter in late January and late November, with Saturn in mid-February and mid-December and with Mercury in mid-April, late September and late October to early November; on all occasions Venus is the brighter object.

Mercury with Mars from mid-June to mid-July when Mercury is the brighter object.

PLANET / STAR CONFUSION TABLE, 2019

Star	Planet	Approximate dates
Antares	Jupiter	January 1 – February 5
Nunki	Saturn	January 19 – December 20
Antares	Venus	January 12 – January 22
Nunki	Venus	February 8 – February 22
Hamal	Mars	February 22 – March 3
Aldebaran	Mars	April 4 – April 27
Pollux	Mars	June 9 – July 2
Aldebaran	Venus	June 16 – June 24
Antares	Jupiter	June 18 – October 5
Spica	Venus	September 27 – October 2
Spica	Mars	November 2 – November 24
Antares	Venus	November 3 – November 17
Nunki	Venus	November 29 – December 14

Figure 4-4. *The 2019 "Do Not Confuse" notes from the NA (top) and from the AA (bottom).*

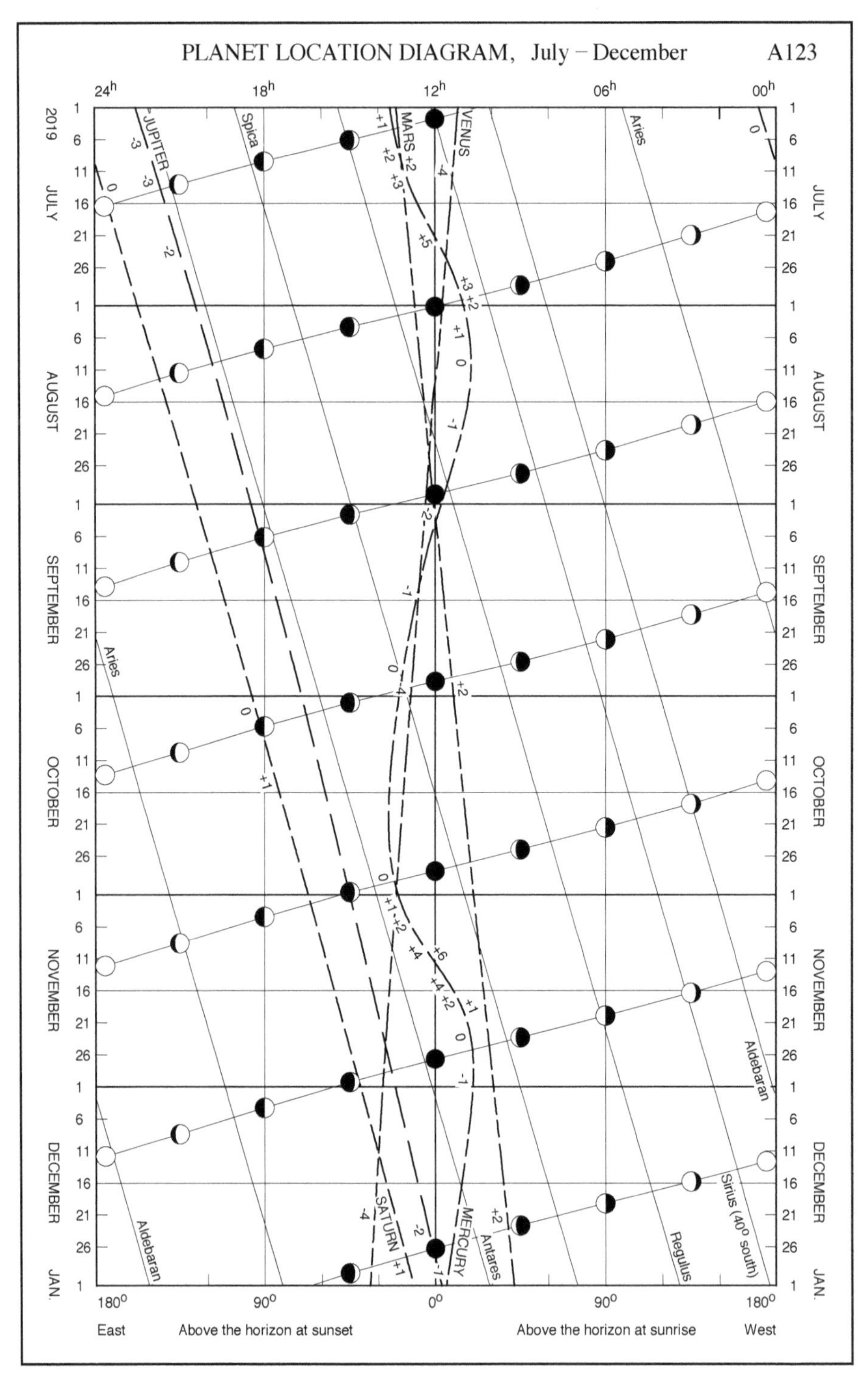

Figure 4-5. *Planet diagram from the AA. Page A122 covers January to June.*

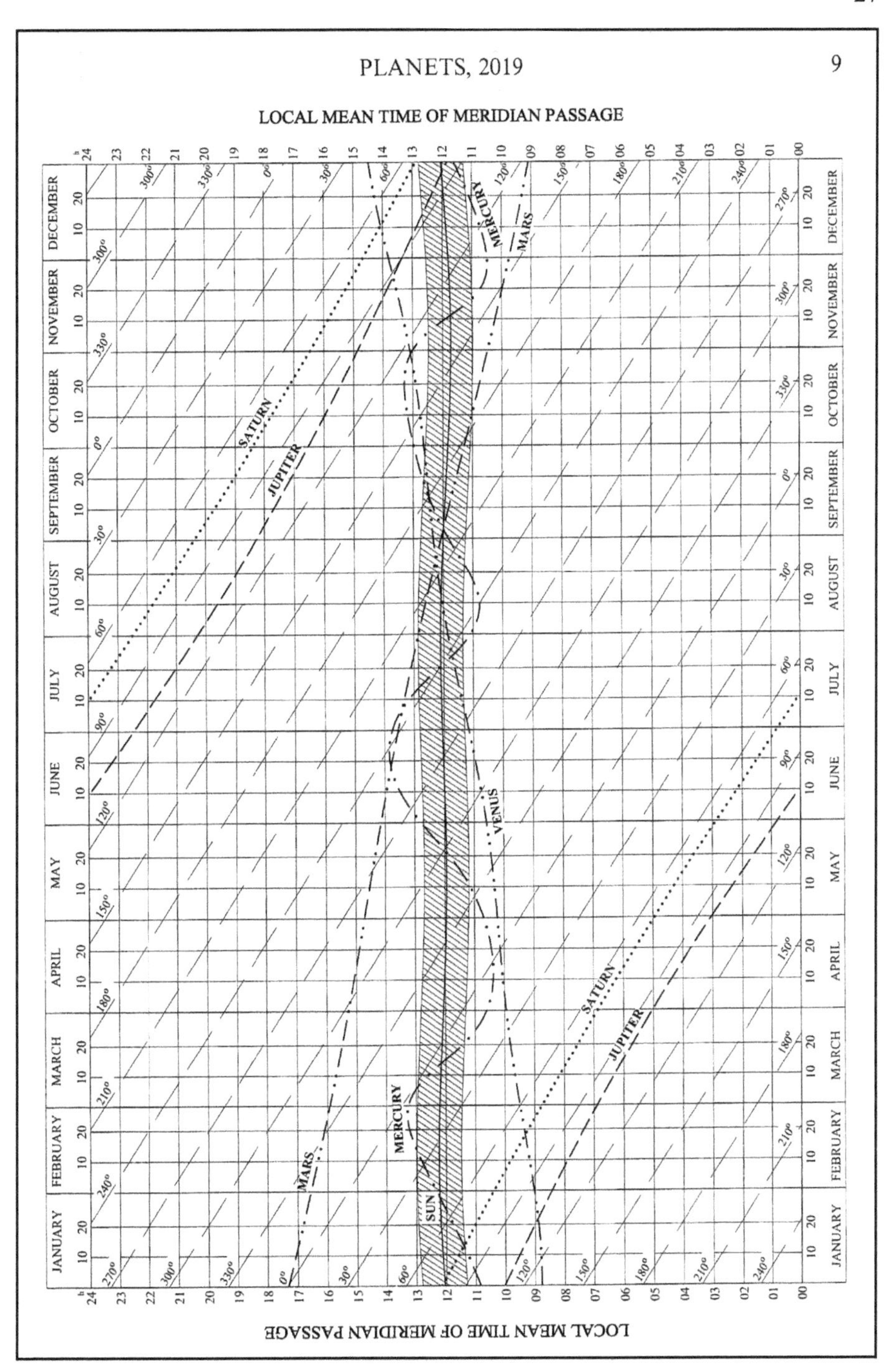

Figure 4-6. *2019 Planet diagram from the NA. These are different each year. NA is one page; AA version is on two pages.*

4.4 Sky Diagrams for Sight Planning

Now we come to the Sky Diagrams, a unique and valuable resource of the AA that we describe here briefly, and then direct those interested to the in-depth discussion of their use in Appendix A1.

These diagrams are used to choose the best sextant sights for a specific sight session at sea, combining both stars and planets during twilight, as well as for finding the best times for sun-moon fixes during the day. They are meant to be used by inspection alone, without special computations or references to other materials.

This method has advantages over Pub 249, Vol. 1, which recommends three stars to use, but does not include planets, and does require almanac data. Also in principle it could be better than using the 2102-D Star Finder, which can be set up to include both stars and planets, but it takes prep work to do so, and it too requires almanac look-ups. There are a set of diagrams for the 15th of each month, organized as Morning, Day, and Evening Sky. A sample is shown in Figure 4-7, illustrated with the observer's vessel heading.

We see that at 25N on July 15 at 17h LMT, the sun is just about to set to the north of west, with Venus (V) preceding it—thus it will be a morning star the next day—and Mars (M) following it over the horizon as an evening star. The moon is not visible. Two hours later at 19h, the day-15 moon is likely on the beam, low on the horizon, with Mars being an evening star about 25° high, located 20° north of west. The star Regulus (#26) is due west, about 30° high, broad on our starboard bow. And so on...

With this type of presentation, we can apply the basic principles (location, height, and brightness) to choose the best triad of bodies for the evening sights. It will not matter that we are not there at precisely 17h or 19h LMT, nor that we are not there on precisely July 15, because the bodies will not have moved much, relative to our criteria for choosing the optimum triad. (These points are demonstrated in Appendix A1.) The arrows in the 17h plot show how much the stars move during 15 days, before or after July 15.

The Sky Diagrams take up 71 pages of the AA. There are another 23 pages of *Polar Sky Diagrams* that are intended for navigation between 75N and 90N. This is common ground for aircraft, but not marine craft. These do not have application to typical ocean navigation, but these views of the sky when standing at the North Pole are interesting on their own—our planetarium talks on star ID, for example, always start at the North Pole.

The Sky Diagrams are a unique part of the AA, in that even if a mariner chose to use the NA for daily work, they would still want a set of these diagrams for sextant sight planning. To that end, we include a detailed "user's guide" to these diagrams in Appendix A1, because the instructions in the AA are relatively brief compared to the effectiveness of this method.

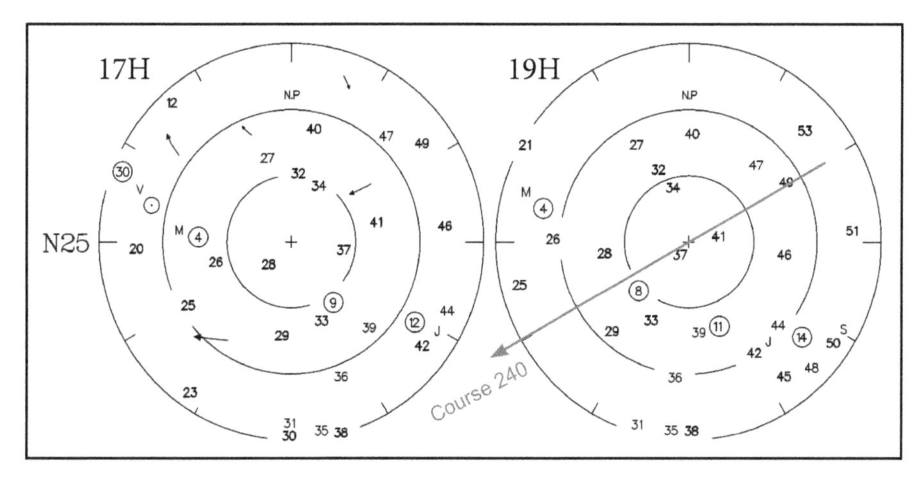

Figure 4-7. *Samples of the Evening Sky for July 15, 2019 at 1700 and 1900 LMT. North is at the top of the diagram; the circumference is the horizon; the center is overhead; each ring is 30° of altitude. Positions of the navigational stars are marked with their unique numbers; the letters are the planets; circle with center dot is the sun; circles with a date inside mark the moon position on those dates; NP is the north pole of the sky, which will always be at a height equal to our Lat; in this case 25°N. The vessel course of 240 is overlaid to obtain relative directions to the bodies.*

5. MISCELLANEOUS TOPICS

5.1 Day of the Year and World Standard Times

A calendar showing the day number within the year, day of the year (DOY), is more valuable in a long voyage than might be guessed. For any type of navigation it is useful for determining ETAs that span long distances, and when relying on traditional celestial navigation based on watch time and watch error, we use it to compute the current watch error relative to when it was set.

The AA shows this DOY on every daily page (see Figure 2-1), whereas the NA puts this information into a Calendar Table on page 5, along with holiday and civil calendar notes on page 4. The AA does not include these extra notes on special days of the year.

Both the AA and the NA have a long list of the nations and places with their standard time zones given, along with notes on daylight savings time at each. These tables are identical in the AA (pages A20 to A23) and the NA (pages 262 to 265). These are useful for arrival planning and shore-side communications.

5.2 Altitude Corrections (Dip, Refraction, Semi-diameter)

Dip

Navigators look to the almanac for corrections to sextant measurements as part of the sight reduction process, and this is an area where we would be right to suspect that there are differences. Sights from flight altitudes using aircraft sextants are different from those using a marine sextant on the deck of a vessel at sea. An obvious difference is aircraft sextants typically are a bubble type that do not rely on a view of the horizon, whereas marine sights use the visible horizon, which dips below the true horizontal as the height of eye (HE) is increases. This makes all sextant angles slightly too large, calling for a *dip correction*. The AA is aware of potential marine use of the almanac, so they include on the last page a short table called Corrections to be Applied to Marine Sextant Observations, shown here in Figure 5-1.

We see that they must be imagining some very large ships, as the height of eye in that table goes to over 2,000 ft! In practice, most marine sights would be from the first column. These corrections are rounded to the nearest whole minute. The counterpart of this table in the NA gives the correction to the tenth. It is part of the Altitude Correction Table on page A2, which in the NA is on the inside front cover and on one side of the yellow bookmark. The NA version for heights above 10° is shown in Figure 5-2. There is another table for the heights below 10° because the refraction is larger and changing more quickly at these lower sextant heights.

CORRECTIONS TO BE APPLIED
TO MARINE SEXTANT ALTITUDES

CORRECTION FOR DIP OF THE HORIZON
To be subtracted from sextant altitude.

Ht.	Dip	Ht.	Dip	Ht.	Dip	Ht.	Dip	Ht.	Dip
Ft.	′	Ft.	′	Ft.	′	Ft.	′	Ft.	′
0		114		437		968		1 707	
	1		11		21		31		41
2		137		481		1 033		1 792	
	2		12		22		32		42
6		162		527		1 099		1 880	
	3		13		23		33		43
12		189		575		1 168		1 970	
	4		14		24		34		44
21		218		625		1 239		2 061	
	5		15		25		35		45
31		250		677		1 311		2 155	
	6		16		26		36		46
43		283		731		1 386		2 251	
	7		17		27		37		47
58		318		787		1 463		2 349	
	8		18		28		38		48
75		356		845		1 543		2 449	
	9		19		29		39		49
93		395		906		1 624		2 551	
	10		20		30		40		50
114		437		968		1 707		2 655	

CORRECTIONS

In addition to sextant error and dip, corrections are to be applied for:
- Refraction
- Semi-diameter (for the Sun and Moon)
- Parallax (for the Moon)
- Dome refraction (if applicable)

MARINE SEXTANT ERROR

Sextant Number

Index Error

Figure 5-1. (Left) *The AA version of the dip correction table. Ht is the height read from the sextant, called Hs in marine navigation.*

Figure 5-2. (Right) *The NA version of the dip correction table, being part of the Altitude Corrections Table. The NA version includes both feet and meters for the height of eye. The AA version is feet only. The stars and planets correction (left column) is pure refraction. The small additional corrections for planets (right column) vary from year to year, but the rest of the table is valid for any year.*

A2 ALTITUDE CORRECTION TABLES 10°-90°—SUN, STARS, PLANETS

SUN

OCT.—MAR. App. Alt.	Lower Limb	Upper Limb	APR.—SEPT. App. Alt.	Lower Limb	Upper Limb
° ′	′	′	° ′	′	′
9 33	+10·8	−21·5	9 39	+10·6	−21·2
9 45	+10·9	−21·4	9 50	+10·7	−21·1
9 56	+11·0	−21·3	10 02	+10·8	−21·0
10 08	+11·1	−21·2	10 14	+10·9	−20·9
10 20	+11·2	−21·1	10 27	+11·0	−20·8
10 33	+11·3	−21·0	10 40	+11·1	−20·7
10 46	+11·4	−20·9	10 53	+11·2	−20·6
11 00	+11·5	−20·8	11 07	+11·3	−20·5
11 15	+11·6	−20·7	11 22	+11·4	−20·4
11 30	+11·7	−20·6	11 37	+11·5	−20·3
11 45	+11·8	−20·5	11 53	+11·6	−20·2
12 01	+11·9	−20·4	12 10	+11·7	−20·1
12 18	+12·0	−20·3	12 27	+11·8	−20·0
12 36	+12·1	−20·2	12 45	+11·9	−19·9
12 54	+12·2	−20·1	13 04	+12·0	−19·8
13 14	+12·3	−20·0	13 24	+12·1	−19·7
13 34	+12·4	−19·9	13 44	+12·2	−19·6
13 55	+12·5	−19·8	14 06	+12·3	−19·5
14 17	+12·6	−19·7	14 29	+12·4	−19·4
14 41	+12·7	−19·6	14 53	+12·5	−19·3
15 05	+12·8	−19·5	15 18	+12·6	−19·2
15 31	+12·9	−19·4	15 45	+12·7	−19·1
15 59	+13·0	−19·3	16 13	+12·8	−19·0
16 27	+13·1	−19·2	16 43	+12·9	−18·9
16 58	+13·2	−19·1	17 14	+13·0	−18·8
17 30	+13·3	−19·0	17 47	+13·1	−18·7
18 05	+13·4	−18·9	18 23	+13·2	−18·6
18 41	+13·5	−18·8	19 00	+13·3	−18·5
19 20	+13·6	−18·7	19 41	+13·4	−18·4
20 02	+13·7	−18·6	20 24	+13·5	−18·3
20 46	+13·8	−18·5	21 10	+13·6	−18·2
21 34	+13·9	−18·4	21 59	+13·7	−18·1
22 25	+14·0	−18·3	22 52	+13·8	−18·0
23 20	+14·1	−18·2	23 49	+13·9	−17·9
24 20	+14·2	−18·1	24 51	+14·0	−17·8
25 24	+14·3	−18·0	25 58	+14·1	−17·7
26 34	+14·4	−17·9	27 11	+14·2	−17·6
27 50	+14·5	−17·8	28 31	+14·3	−17·5
29 13	+14·6	−17·7	29 58	+14·4	−17·4
30 44	+14·7	−17·6	31 33	+14·5	−17·3
32 24	+14·8	−17·5	33 18	+14·6	−17·2
34 15	+14·9	−17·4	35 15	+14·7	−17·1
36 17	+15·0	−17·3	37 24	+14·8	−17·0
38 34	+15·1	−17·2	39 48	+14·9	−16·9
41 06	+15·2	−17·1	42 28	+15·0	−16·8
43 56	+15·3	−17·0	45 29	+15·1	−16·7
47 07	+15·4	−16·9	48 52	+15·2	−16·6
50 43	+15·5	−16·8	52 41	+15·3	−16·5
54 46	+15·6	−16·7	56 59	+15·4	−16·4
59 21	+15·7	−16·6	61 50	+15·5	−16·3
64 28	+15·8	−16·5	67 15	+15·6	−16·2
70 10	+15·9	−16·4	73 14	+15·7	−16·1
76 24	+16·0	−16·3	79 42	+15·8	−16·0
83 05	+16·1	−16·2	86 31	+15·9	−15·9
90 00			90 00		

STARS AND PLANETS

App Alt.	Corr^n
° ′	′
9 55	−5·3
10 07	−5·2
10 20	−5·1
10 32	−5·0
10 46	−4·9
10 59	−4·8
11 14	−4·7
11 29	−4·6
11 44	−4·5
12 00	−4·4
12 17	−4·3
12 35	−4·2
12 53	−4·1
13 12	−4·0
13 32	−3·9
13 53	−3·8
14 16	−3·7
14 39	−3·6
15 03	−3·5
15 29	−3·4
15 56	−3·3
16 25	−3·2
16 55	−3·1
17 27	−3·0
18 01	−2·9
18 37	−2·8
19 16	−2·7
19 56	−2·6
20 40	−2·5
21 27	−2·4
22 17	−2·3
23 11	−2·2
24 09	−2·1
25 12	−2·0
26 20	−1·9
27 34	−1·8
28 54	−1·7
30 22	−1·6
31 58	−1·5
33 43	−1·4
35 38	−1·3
37 45	−1·2
40 06	−1·1
42 42	−1·0
45 34	−0·9
48 45	−0·8
52 16	−0·7
56 09	−0·6
60 26	−0·5
65 06	−0·4
70 09	−0·3
75 32	−0·2
81 12	−0·1
87 03	0·0
90 00	

App. Alt.	Additional Corr^n
2019	
VENUS	
Jan. 1–Feb. 15	
°	′
0	+0·2
41	+0·1
76	
Feb. 16–Dec. 31	
°	′
0	+0·1
60	
MARS	
Jan. 1–Dec. 31	
°	′
0	+0·1
60	

DIP

Ht. of Eye (m)	Corr^n (′)	Ht. of Eye (ft.)
2·4	−2·8	8·0
2·6	−2·9	8·6
2·8	−3·0	9·2
3·0	−3·1	9·8
3·2	−3·2	10·5
3·4	−3·3	11·2
3·6	−3·4	11·9
3·8	−3·5	12·6
4·0	−3·6	13·3
4·3	−3·7	14·1
4·5	−3·8	14·9
4·7	−3·9	15·7
5·0	−4·0	16·5
5·2	−4·1	17·4
5·5	−4·2	18·3
5·8	−4·3	19·1
6·1	−4·4	20·1
6·3	−4·5	21·0
6·6	−4·6	22·0
6·9	−4·7	22·9
7·2	−4·8	23·9
7·5	−4·9	24·9
7·9	−5·0	26·0
8·2	−5·1	27·1
8·5	−5·2	28·1
8·8	−5·3	29·2
9·2	−5·4	30·4
9·5	−5·5	31·5
9·9	−5·6	32·7
10·3	−5·7	33·9
10·6	−5·8	35·1
11·0	−5·9	36·3
11·4	−6·0	37·6
11·8	−6·1	38·9
12·2	−6·2	40·1
12·6	−6·3	41·5
13·0	−6·4	42·8
13·4	−6·5	44·2
13·8	−6·6	45·5
14·2	−6·7	46·9
14·7	−6·8	48·4
15·1	−6·9	49·8
15·5	−7·0	51·3
16·0	−7·1	52·8
16·5	−7·2	54·3
16·9	−7·3	55·8
17·4	−7·4	57·4
17·9	−7·5	58·9
18·4	−7·6	60·5
18·8	−7·7	62·1
19·3	−7·8	63·8
19·8	−7·9	65·4
20·4	−8·0	67·1
20·9	−8·1	68·8
21·4		70·5

Ht. of Eye	Corr^n
m	′
1·0	− 1·8
1·5	− 2·2
2·0	− 2·5
2·5	− 2·8
3·0	− 3·0
See table ←	
m	′
20	− 7·9
22	− 8·3
24	− 8·6
26	− 9·0
28	− 9·3
30	− 9·6
32	−10·0
34	−10·3
36	−10·6
38	−10·8
40	−11·1
42	−11·4
44	−11·7
46	−11·9
48	−12·2
ft.	′
2	− 1·4
4	− 1·9
6	− 2·4
8	− 2·7
10	− 3·1
See table ←	
ft.	′
70	− 8·1
75	− 8·4
80	− 8·7
85	− 8·9
90	− 9·2
95	− 9·5
100	− 9·7
105	− 9·9
110	−10·2
115	−10·4
120	−10·6
125	−10·8
130	−11·1
135	−11·3
140	−11·5
145	−11·7
150	−11·9
155	−12·1

App. Alt. = Apparent altitude = Sextant altitude corrected for index error and dip.

The NA version is more precise, but if needed this can be computed as:

Dip correction = 0.97' Sqrt[HE(ft)] = 1.76' Sqrt[HE(m)].

Refraction

When light rays from celestial bodies enter the atmosphere from the vacuum of space, they bend down toward the earth, which means the altitude we observe with a sextant is slightly too high. We fix this with the *refraction correction*. It is largest for very low sights and diminishes rapidly for sights above 20° or so. The AA version of this is a complex table on page A167 oriented toward aircraft sights. It is reproduced in Figure 5-3.

The NA includes the refraction corrections indirectly in the Altitude Correction Table A2, Figure 5-2. It is the column headed "Stars and Planets," because for these bodies the total altitude correction is made up of refraction. Thus if you were using AA for most other data, then this is where you would get superior corrections for refraction. For example, the AA table (Figure 5-3) gives a correction of -4' for all altitudes between 12° and 16°, whereas we see in the NA table (Stars and Planets Column of Figure 5-2) that these corrections in that altitude range vary from -4.4' to -3.3'. These differences are all within the 1' accuracy discussed in the AA, but there is no reason to throw away accuracy when we do not need to. Even using the AA for most other data, we can keep a copy of the altitude corrections from the NA to refer to. These refraction corrections do not change from year to year, so a copy from any expired NA will do the job.

In the bottom of Figure 5-3 we see the AA corrections to refraction caused by abnormal temperatures. These are approximations to the values included in the NA in Table A4. These extra corrections also depend on atmospheric pressure, but that is not addressed in the AA. These extra corrections are only important for low sights (below 10°) and in extreme conditions.

Semi-diameter

A similar situation occurs with the sight reduction of sun and moon sights. Sun sights are usually taken to the lower limb; moon sights can be upper or lower. For sight reduction and comparison with almanac data, however, we need to know the measurement to the center of the body, which means applying a semi-diameter correction. In the AA, the semi-diameter (SD) is on each daily page. See, for example, Figure 2-1. The sun's value is accurate to the tenth of a minute; the moon's only accurate to the whole minute. These are both accurate to the tenth in the NA, but again we stress that this has little practical implication as it is not often we can expect moon sights more accurate than about ±0.5'.

Furthermore, for marine application of the AA, it is simply best to keep a copy of the NA altitude correction tables at hand and use them. These corrections will then be easier to make and more accurate.

CORRECTIONS TO BE APPLIED TO SEXTANT ALTITUDE A167

REFRACTION

To be subtracted from sextant altitude (referred to as observed altitude in A.P. 3270)

R_o	0	5	10	15	20	25	30	35	40	45	50	55	R_o	0·9	1·0	1·1	1·2
	Height above sea level in units of 1 000 ft. — Sextant Altitude													$R = R_o \times f$ — f / R			
′	° ′	° ′	° ′	° ′	° ′	° ′	° ′	° ′	° ′	° ′	° ′	° ′	′	′	′	′	′
	90	90	90	90	90	90	90	90	90	90	90	90					
0													0	0	0	0	0
	63	59	55	51	46	41	36	31	26	20	17	13					
1													1	1	1	1	1
	33	29	26	22	19	16	14	11	9	7	6	4					
2													2	2	2	2	2
	21	19	16	14	12	10	8	7	5	4	2 40	1 40					
3													3	3	3	3	4
	16	14	12	10	8	7	6	5	3 10	2 20	1 30	0 40					
4													4	4	4	4	5
	12	11	9	8	7	5	4 00	3 10	2 10	1 30	0 39	+0 05					
5													5	5	5	5	6
	10	9	7	5 50	4 50	3 50	3 10	2 20	1 30	0 49	+0 11	−0 19					
6													6	5	6	7	7
	8 10	6 50	5 50	4 50	4 00	3 00	2 20	1 50	1 10	0 24	−0 11	−0 38					
7													7	6	7	8	8
	6 50	5 50	5 00	4 00	3 10	2 30	1 50	1 20	0 38	+0 04	−0 28	−0 54					
8													8	7	8	9	10
	6 00	5 10	4 10	3 20	2 40	2 00	1 30	1 00	0 19	−0 13	−0 42	−1 08					
9													9	8	9	10	11
	5 20	4 30	3 40	2 50	2 10	1 40	1 10	0 35	+0 03	−0 27	−0 53	−1 18					
10													10	9	10	11	12
	4 30	3 40	2 50	2 20	1 40	1 10	0 37	+0 11	−0 16	−0 43	−1 08	−1 31					
12													12	11	12	13	14
	3 30	2 50	2 10	1 40	1 10	0 34	+0 09	−0 14	−0 37	−1 00	−1 23	−1 44					
14													14	13	14	15	17
	2 50	2 10	1 40	1 10	0 37	+0 10	−0 13	−0 34	−0 53	−1 14	−1 35	−1 56					
16													16	14	16	18	19
	2 20	1 40	1 20	0 43	+0 15	−0 08	−0 31	−0 52	−1 08	−1 27	−1 46	−2 05					
18													18	16	18	20	22
	1 50	1 20	0 49	+0 23	−0 02	−0 26	−0 46	−1 06	−1 22	−1 39	−1 57	−2 14					
20													20	18	20	22	24
	1 12	0 44	+0 19	−0 06	−0 28	−0 48	−1 09	−1 27	−1 42	−1 58	−2 14	−2 30					
25													25	22	25	28	30
	0 34	+0 10	−0 13	−0 36	−0 55	−1 14	−1 32	−1 51	−2 06	−2 21	−2 34	−2 49					
30													30	27	30	33	36
	+0 06	−0 16	−0 37	−0 59	−1 17	−1 33	−1 51	−2 07	−2 23	−2 37	−2 51	−3 04					
35													35	31	35	38	42
	−0 18	−0 37	−0 58	−1 16	−1 34	−1 49	−2 06	−2 22	−2 35	−2 49	−3 03	−3 16					
40													40	36	40	44	48
		−0 53	−1 14	−1 31	−1 47	−2 03	−2 18	−2 33	−2 47	−2 59	−3 13	−3 25					
45													45	40	45	50	54
		−1 10	−1 28	−1 44	−1 59	−2 15	−2 28	−2 43	−2 56	−3 08	−3 22	−3 33					
50													50	45	50	55	60
			−1 40	−1 53	−2 09	−2 24	−2 38	−2 52	−3 04	−3 17	−3 29	−3 41					
55													55	49	55	60	66
				−2 03	−2 18	−2 33	−2 46	−3 01	−3 12	−3 25	−3 37	−3 48					
60													60	54	60	66	72
							−2 53	−3 07	−3 19	−3 31	−3 42	−3 53					

f	0	5	10	15	20	25	30	35	40	45	50	55	f
	Temperature in °C.												
	+47	+36	+27	+18	+10	+ 3	− 5	−13					
0·9													0·9
	+26	+16	+ 6	− 4	−13	−22	−31	−40					
1·0													1·0
	+ 5	− 5	−15	−25	−36	−46	−57	−68					
1·1													1·1
	−16	−25	−36	−46	−58	−71	−83	−95					
1·2													1·2
	−37	−45	−56	−67	−81	−95							

For these heights no temperature correction is necessary, so use $R = R_o$

0·9 1·0 1·1 1·2 — f: Where R_o is less than 10′ or the height greater than 35 000 ft. use $R = R_o$

Choose the column appropriate to height, in units of 1 000 ft., and find the range of altitude in which the sextant altitude lies; the corresponding value of R_o is the refraction, to be subtracted from sextant altitude, unless conditions are extreme. In that case find f from the lower table, with critical argument temperature. Use the table on the right to form the refraction, $R = R_o \times f$.

Figure 5-3. *The AA version of refraction corrections. Marine applications on the surface would all be from the left-hand columns we have marked in boxes. Values can be compared with those in the Stars and Planets column of Figure 5-2.*

5.3 Latitude by Polaris

A workhorse of celestial navigation in the Northern Hemisphere is the process of finding latitude from the sextant height of Polaris. Like finding Latitude from noon sights, its popularity can be traced to its simplicity. Just a few corrections are needed, without any sight reduction tables.

The procedure used in the AA is different from that used in the NA, but the results are close if we modify the process a bit. The AA method has fewer steps with only a very slight sacrifice in accuracy. The Polaris correction table is on page A157 (Figure 5-4). We enter the table with Local Hour Angle of Aries:

LHA Aries = GHA Aries - Lon W (or + Lon E). Here is an example.

DR position 31 N, 140 W, evening twilight on Mar 1, 2019, UTC = 04:04:55, sextant height of Polaris is Hs = 31° 4.5'. Index correction of the sextant is 0; the height of eye of the observer at the time was 10 ft. Find Latitude.

First find GHA Aries (daily pages, Figure 2-1) for 4h 0m = 218° 40.5'. Then get correction for 4m 55s from page A165, which is 1° 14.0', so GHA Aries = 219° 54.5'. From this we find LHA Aries = 219° 54.5' – 140 = 79° 54.5'. Using this we enter Polaris table, where we find the answer between 78 05 and 80 35 to be Q = –32'.

At this point, the AA Explanation tells us to subtract the refraction from Hs and apply the Q correction from the Polaris tables. But we know we have a dip correction to make, and we know the AA refraction tables are not as precise as the NA values, so we will modify this procedure and use the NA corrections in Figure 5-2. We also note that Hs corrected for index error, dip, and refraction is called Ho, the observed altitude. Thus we have in the AA format:

$$\begin{aligned} \text{Lat} &= \text{Hs} - \text{Dip} - \text{Refraction} \pm Q \\ &= \text{Ho} + Q \\ &= 31^\circ\ 4.5' - 3.1' - 1.6' + Q' \\ &= 31^\circ\ 0.2' + Q \\ &= 31^\circ\ 0.2' - 32' = 30^\circ\ 28.2'\ \text{N}. \end{aligned}$$

This result can be compared to what we get from the NA, which does not use a single Q correction, but rather uses this formula:

$$\text{Lat} = \text{Ho} - 1^\circ + a0 + a1 + a2,$$

where the three corrections are from the NA version of the Polaris table. The -1° is there so that the remaining corrections in the NA can all be tabulated as positive. The a0 correction depends only on LHA Aries and is 27.1'; the a1 depends on Lat and at 30° is 0.6'; and the a2 depends on the month. For March it is 0.8'. So the total NA correction is –1° + 27.1' +0.6' + 0.8' = –31.5', which is the equivalent of the Q value used in the AA. Hence we get from the NA:

$$\text{Lat} = \text{Ho} + (-1^\circ + a0 + a1 + a2) = 31^\circ\ 0.2' - 31.5' = 30^\circ\ 28.7'\ \text{N}.$$

Using the NA correction tables reduces the discrepancy in the AA value to the differences in the total correction: AA = –32; NA = – 31.5. This difference is well within the stated limits of the AA and indeed of practical cel nav in general. It seems reasonable then to use the shorter Q table from the AA, but use the NA for altitude corrections.

The single Q-value correction for Polaris sights is the method we do in our booklet *GPS Backup with a Mark 3 Sextant*. The Polaris table in that book uses a universal procedure for finding Q from the relative positions of stars on either side of Polaris that can be applied on any date, at any north latitude.

The azimuth of Polaris (000 ± 1.5° in most cases) is used to check gyro compasses. Both almanacs include a table of these values as a function of LHA Aries and Lat. The two tables yield the same results.

POLARIS (POLE STAR) TABLE, 2019 A157

FOR DETERMINING THE LATITUDE FROM A SEXTANT ALTITUDE

LHA Aries	Q	LHA Aries	Q	LHA Aries	Q	LHA Aries	Q	LHA Aries	Q	LHA Aries	Q	LHA Aries	Q	LHA Aries	Q
° ′	′	° ′	′	° ′	′	° ′	′	° ′	′	° ′	′	° ′	′	° ′	′
358 42	−28	**87 20**	−28	**124 12**	− 6	**157 00**	+16	**206 34**	+38	**286 15**	+18	**319 30**	− 4	**354 46**	−26
0 47	−29	**89 25**	−27	**125 41**	− 5	**158 36**	+17	**212 25**	+39	**287 54**	+17	**320 58**	− 5	**356 42**	−27
2 56	−30	**91 25**	−26	**127 09**	− 4	**160 13**	+18	**235 42**	+38	**289 31**	+16	**322 26**	− 6	**358 42**	−28
5 11	−31	**93 21**	−25	**128 37**	− 3	**161 52**	+19	**241 33**	+37	**291 07**	+15	**323 55**	− 7	**0 47**	−29
7 32	−32	**95 14**	−24	**130 05**	− 2	**163 32**	+20	**245 56**	+36	**292 42**	+14	**325 23**	− 8	**2 56**	−30
10 02	−33	**97 04**	−23	**131 32**	− 1	**165 14**	+21	**249 36**	+35	**294 16**	+13	**326 53**	− 9	**5 11**	−31
12 43	−34	**98 51**	−22	**133 00**	0	**166 57**	+22	**252 49**	+34	**295 49**	+12	**328 22**	−10	**7 32**	−32
15 36	−35	**100 37**	−21	**134 28**	+ 1	**168 43**	+23	**255 45**	+33	**297 21**	+11	**329 53**	−11	**10 02**	−33
18 48	−36	**102 20**	−20	**135 56**	+ 2	**170 32**	+24	**258 27**	+32	**298 52**	+10	**331 24**	−12	**12 43**	−34
22 26	−37	**104 01**	−19	**137 23**	+ 3	**172 22**	+25	**260 58**	+31	**300 23**	+ 9	**332 55**	−13	**15 36**	−35
26 45	−38	**105 41**	−18	**138 51**	+ 4	**174 16**	+26	**263 21**	+30	**301 53**	+ 8	**334 28**	−14	**18 48**	−36
32 33	−39	**107 19**	−17	**140 19**	+ 5	**176 13**	+27	**265 37**	+29	**303 22**	+ 7	**336 01**	−15	**22 26**	−37
55 34	−38	**108 56**	−16	**141 47**	+ 6	**178 14**	+28	**267 48**	+28	**304 51**	+ 6	**337 36**	−16	**26 45**	−38
61 22	−37	**110 31**	−15	**143 16**	+ 7	**180 19**	+29	**269 53**	+27	**306 20**	+ 5	**339 11**	−17	**32 33**	−39
65 41	−36	**112 06**	−14	**144 45**	+ 8	**182 30**	+30	**271 54**	+26	**307 48**	+ 4	**340 48**	−18	**55 34**	−38
69 19	−35	**113 39**	−13	**146 14**	+ 9	**184 46**	+31	**273 51**	+25	**309 16**	+ 3	**342 26**	−19	**61 22**	−37
72 31	−34	**115 12**	−12	**147 44**	+10	**187 09**	+32	**275 45**	+24	**310 44**	+ 2	**344 06**	−20	**65 41**	−36
75 24	−33	**116 43**	−11	**149 15**	+11	**189 40**	+33	**277 35**	+23	**312 11**	+ 1	**345 47**	−21	**69 19**	−35
78 05	−32	**118 14**	−10	**150 46**	+12	**192 22**	+34	**279 24**	+22	**313 39**	0	**347 30**	−22	**72 31**	−34
80 35	−31	**119 45**	− 9	**152 18**	+13	**195 18**	+35	**281 10**	+21	**315 07**	− 1	**349 16**	−23	**75 24**	−33
82 56	−30	**121 14**	− 8	**153 51**	+14	**198 31**	+36	**282 53**	+20	**316 35**	− 2	**351 03**	−24	**78 05**	−32
85 11	−29	**122 44**	− 7	**155 25**	+15	**202 11**	+37	**284 35**	+19	**318 02**	− 3	**352 53**	−25	**80 35**	−31
87 20		**124 12**		**157 00**		**206 34**		**286 15**		**319 30**		**354 46**		**82 56**	

In critical cases, ascend

Q, which does not include refraction, is to be applied to the corrected sextant altitude of *Polaris*.

Polaris: Mag. 2.0, SHA 315° 56′, Dec N 89° 20′.7

Figure 5-4. *The AA version of the Polaris correction table used to find latitude from the sextant height of Polaris.*

5.4 Sight Reduction Tables

A big difference between the two documents, AA and NA, is not related at all to the almanac content of either, which we have seen here is in practice equivalent, but rather to something else related to celestial navigation. Starting in the year 1989, the USNO included in each printing of the NA a complete copy of a set of sight reduction tables. They are referred to as "concise" tables, meaning they are much shorter (34 pages) than the Pub 229 or 249 versions, which are large, multiple volume books, but the cost of this conciseness is several extra steps in getting the answers. The tables are often called the *NAO Sight Reduction Table*s, reflecting their source in the Nautical Almanac Office.

With these tables included, the NA becomes a complete one book solution to celestial navigation position fixing. When we use the AA for almanac data, we must use an additional set of books for the sight reduction. To many navigators this is a non issue as they do not like the extra steps needed in the NAO Tables, but to others this could be a factor. For those who count on doing cel nav underway, but plan to do it primary with a calculator or computer program, the NA is a convenient one-book backup that they do not plan to use except in a contingency.

6. SUMMARY

The AA provides essentially all the standard almanac data a mariner expects from the NA. The precision of sun and stars data (0.1') is identical; the moon and planet data in the AA are given only to the nearest whole minute. The overall stated accuracy of AA data is slightly less than that of the NA, but this is not a factor in most applications, where the accuracy of the final position fix underway is more determined by sextant sighting skills and how the motion of the vessel is accounted for during the sight taking session.

This review has also shown why the AA is preferred over the NA by some recreational marine navigators. The AA features that make it attractive include:

- Declination given every 10 minutes removes the need for a further time correction to the declination
- GHA correction for sun and Aries is a bit simpler than in NA
- Declination listed next to time of mer pass saves one almanac entry
- Only useful planets tabulated is a nice heads up in sight planning
- Polaris corrections are simpler with the Q-values
- Sky Diagrams are a huge plus for sight planning
- The publication is free, compared to $30-$60 for the NA
- Some like that it is digital and can be stored as an ebook for easy access, although traditional cel nav preparation would call for printed materials.

In our cel nav classes and reference books, we point out these virtues of the AA, but still recommend to those starting to learn cel nav from scratch that they use the NA. The main reasons for that recommendation have little to do with the almanac data itself. A primary reason is simply the standardization of the subject matter teaching. There is a lot of bookwork and paperwork in cel nav to begin with, and we do not want to complicate that with even more methods. Furthermore, the AA is not really an option for professional mariners who are in effect required to use the NA and Pub 229 for sight reduction, or their British Counterparts, NP314 and NP401.

Our goal of standardized teaching is the same reason we do not make up logical names or abbreviations for various parameters that have traditionally unclear or even misleading names—i.e., zone description is called ZD; zenith distance is called z; azimuth angle is called Z; and azimuth is called Zn, to give a few examples from the Zs.

And then there is the issue of the altitude correction tables. Even if one chooses the AA for almanac data, we are better off using the sextant sight correction tables from the NA. They are more precise and span the parameter ranges better that are used in marine navigation. A copy from any expired NA does the job.

Another important factor in leaning toward the NA is the inclusion of the NAO Sight Reduction tables, which makes the NA a one book solution to cel nav position fixes. This is of interest to those who first learn the traditional methods using books and paper, but then choose to do the daily work with a dedicated app such as the StarPilot calculator, which is much faster and yields more accurate results. For those navigators, they can purchase the NA, double zip-lock bag it up, and stow it for some contingency use, but otherwise do the work and plotting with the calculator.

Also, the NA includes more discussion of the tables themselves, including, for those who want it, analytical solutions that can be used in personal calculators or computers.

With those perspectives noted, for the mariner who does not have a specific concern in standards of teaching, but reasonably cares mostly about their own safe, efficient navigation, we hope we have shown that the free Air Almanac, along with a copy of sight reduction tables of their choice (concise or standard) and a set of the NA sextant altitude corrections, can well meet the needs of routine cel nav underway.

A1. Sextant Sight Planning with the Air Almanac's Sky Diagrams

A unique feature of the Air Almanac (AA) is its set of Sky Diagrams intended for planning the optimum sextant sights to take for the best fix. We are typically confronted with a sky full of stars and a planet or two, and the choices we make are crucial to optimum accuracy of the resulting fix.

Here we illustrate that selection process using the Sky Diagrams and compare it with two other methods of sight planning, Pub 249 Vol. 1 and the 2102-D Star Finder. We facilitate this comparison with a jump into the electronic world to use the Best Sights function of the StarPilot app (StarPilotLLC.com) to check what the manual methods find. For reference as we proceed, a set of the 2019 Sky Diagrams can be downloaded here: starpath.com/downloads/2019_Sky_Diagrams.pdf

The Sky Diagrams are radar-like plots of the heights and bearings of all celestial bodies in the sky that we might use for a sextant sight. There is a diagram for every 2 hours of LMT for the 15th of each month of the almanac year, for latitudes 50S, 25S, 00, 25N, 50N, and 75N. They are grouped into Morning Sky (01, 03, 05, and 07, LMT), Daytime Sky (09, 11, 13, 15), and Evening Sky (17, 19, 21, 23). The Daytime Sky is mostly for planning sun-moon sights, but also useful in finding these bodies in a cloudy sky, without resorting to computations. Also, in some conditions, the moon and sky are almost the same color, even without clouds. A sample of the Evening Sky is shown in Figure A-1. Figure A-2 shows the legends explaining the symbols. These are shown on alternating diagram pages.

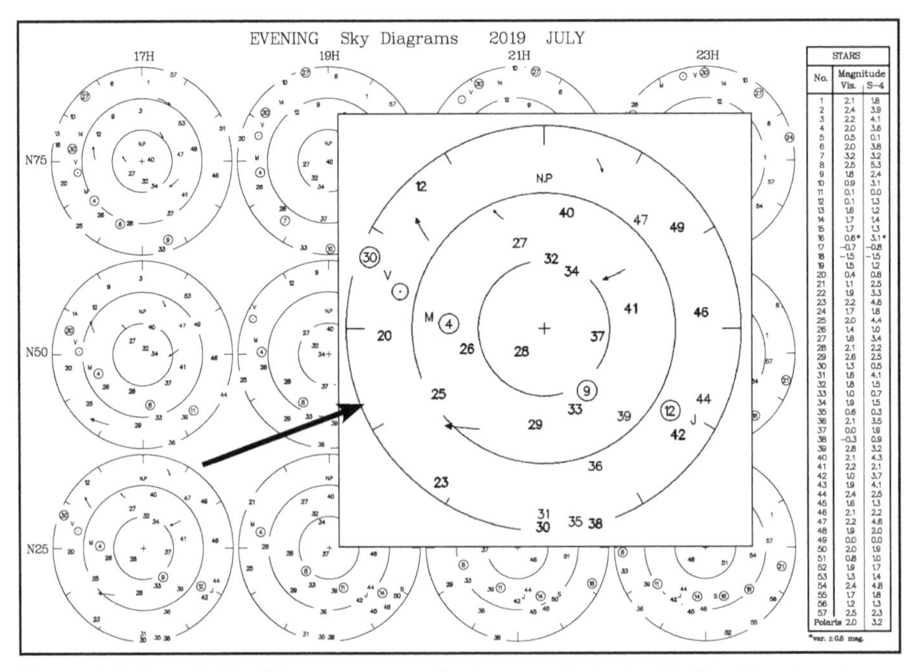

No.	Magnitude Vis.	S–4
1	2.1	1.8
2	2.4	3.9
3	2.2	4.1
4	2.0	3.6
5	0.5	0.1
6	2.0	3.8
7	3.2	3.2
8	2.5	5.3
9	1.8	2.4
10	0.9	3.1
11	0.1	0.0
12	0.1	1.3
13	1.6	1.2
14	1.7	1.4
15	1.7	1.3
16	0.6*	3.1*
17	−0.7	−0.8
18	−1.5	−1.5
19	1.5	1.2
20	0.4	0.8
21	1.1	2.5
22	1.9	3.3
23	2.2	4.6
24	1.7	1.8
25	2.0	4.4
26	1.4	1.0
27	1.8	3.4
28	2.1	2.2
29	2.6	2.5
30	1.3	0.5
31	1.6	4.1
32	1.8	1.5
33	1.0	0.7
34	1.9	1.5
35	0.6	0.3
36	2.1	3.5
37	0.0	1.9
38	−0.3	0.9
39	2.8	3.2
40	2.1	4.3
41	2.2	2.1
42	1.0	3.7
43	1.9	4.1
44	2.4	2.5
45	1.6	1.3
46	2.1	2.2
47	2.2	4.6
48	1.9	2.0
49	0.0	0.0
50	2.0	1.9
51	0.8	1.0
52	1.9	1.7
53	1.3	1.4
54	2.4	4.8
55	1.7	1.8
56	1.2	1.3
57	2.5	2.3
Polaris	2.0	3.2

*var. ±0.5 mag.

Figure A-1. *Sample Sky Diagrams page with star magnitudes beside it, with an insert showing details of one diagram.*

The center of each diagram is overhead; the circumference is the horizon. The rings mark heights above the horizon of 30° and 60°. In the Northern Hemisphere, the north pole of the sky (NP) will be at a height equal to your Lat. The position of the moon is shown several times on each diagram, because it moves east relative to the stars at about 12° per day. Its position on a given date is marked by a circle around that date. The moon position is shown every 3 or 4 days, which we can use to estimate where it is on other days, including the 15th, the base date for the star data.

The arrows on the left-hand diagrams (Figure A-1) show how stars in that vicinity move during the 15 days after the 15th of the month, which can be projected back to show their positions at the first of the month. These can be used for estimating positions on dates other than the 15th, as discussed in the Notes on Star Motions at the end of this section.

When the sun (circle with dot) is showing, the sun is up, and it is daylight. The other bodies would not be visible except for the moon—and sometimes Venus, if it is far enough from the sun. In which case, we can figure from these diagrams exactly where it is located, and then look in the right direction, through the telescope of a pre-set sextant, and perhaps get a daytime Venus sight.

The column headed "S-4" is a relative photo-sensitive response indicating how bright they would appear in a photograph. This is not a useful parameter for us—even though we might, with effort, extract star color information from it; we recommend crossing it out, so it does not confuse anything. Our textbook *Celestial Navigation* and *The Star Finder Book* each have extended discussions of star brightness and the complex visual magnitude scale (see Table A-1).

As an aside, the Sky Diagrams label latitudes with the N in front, such as N25, meaning latitude 25° North. This is unfortunate, because in the marine world of

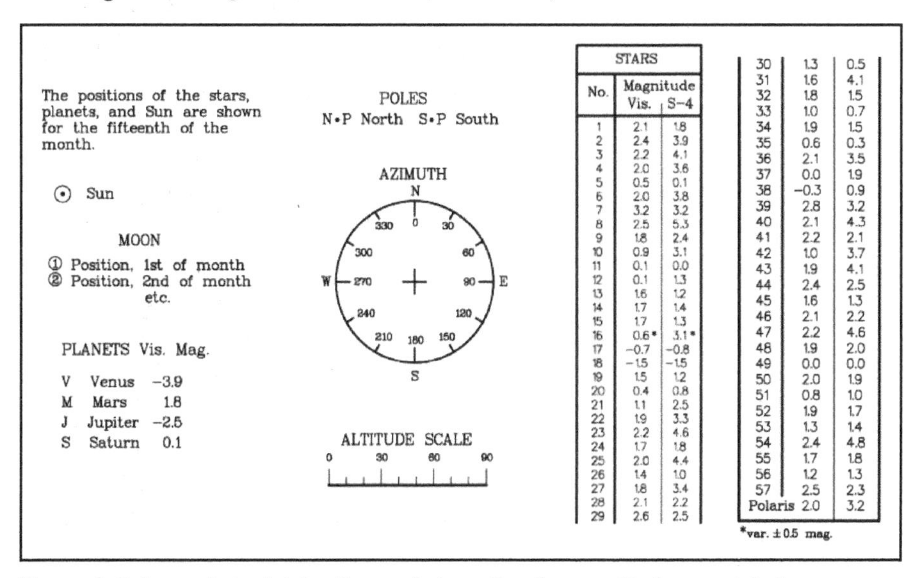

STARS No.	Magnitude Vis.	S-4
1	2.1	1.8
2	2.4	3.9
3	2.2	4.1
4	2.0	3.6
5	0.5	0.1
6	2.0	3.8
7	3.2	3.2
8	2.5	5.3
9	1.8	2.4
10	0.9	3.1
11	0.1	0.0
12	0.1	1.3
13	1.6	1.2
14	1.7	1.4
15	1.7	1.3
16	0.6*	3.1*
17	−0.7	−0.8
18	−1.5	−1.5
19	1.5	1.2
20	0.4	0.8
21	1.1	2.5
22	1.9	3.3
23	2.2	4.6
24	1.7	1.8
25	2.0	4.4
26	1.4	1.0
27	1.8	3.4
28	2.1	2.2
29	2.6	2.5
30	1.3	0.5
31	1.6	4.1
32	1.8	1.5
33	1.0	0.7
34	1.9	1.5
35	0.6	0.3
36	2.1	3.5
37	0.0	1.9
38	−0.3	0.9
39	2.8	3.2
40	2.1	4.3
41	2.2	2.1
42	1.0	3.7
43	1.9	4.1
44	2.4	2.5
45	1.6	1.3
46	2.1	2.2
47	2.2	4.6
48	1.9	2.0
49	0.0	0.0
50	2.0	1.9
51	0.8	1.0
52	1.9	1.7
53	1.3	1.4
54	2.4	4.8
55	1.7	1.8
56	1.2	1.3
57	2.5	2.3
Polaris	2.0	3.2

*var. ±0.5 mag.

Figure A-2. *Legends explaining the symbols on the diagram. Parts are printed on every other page.*

celestial navigation we label latitudes with a following N or S and declinations with a preceding N or S, i.e., the star Arcturus has a declination of about N 19°, which means it circles the earth over latitude 19° N.

These diagrams are best used in print with an overlaid transparency sheet, as shown in Figure A-4, then we go on to compare this method with other methods.

In practice, when we need to predict the best bodies to shoot in a sight session, we are not likely to be at precisely Lat 0°, 25° or 50° (N or S), and indeed for evening sights, the right twilight time is not likely to be exactly 17, 19, 21, or 23 hr LMT. Nevertheless, we will first assume that is the case, to see how we match up with other methods, then we will choose some random set of circumstances such as sights at evening twilight on July 10 at 31N, 140W—an approximate point in time and space occupied by many sailors during any of several trans-pacific ocean races.

We look first at a DR Lat of exactly 25° N on July 15, 2019. In Figure A-1, we see the sun is still up at 17h LMT, so let's use 19h LMT as the test example, which must be fairly close to twilight. Figure A-3 shows that sky. We marked the bright stars with arrows. Jupiter and Mars are also available, as is the moon estimated to be at about 15° above the SE horizon on this July 15 sky, based on the July 14th position shown.

Now that we know what the sky looks like, we need to ask how do we select the best sights? Refer to our books cited earlier for the basis of these criteria.

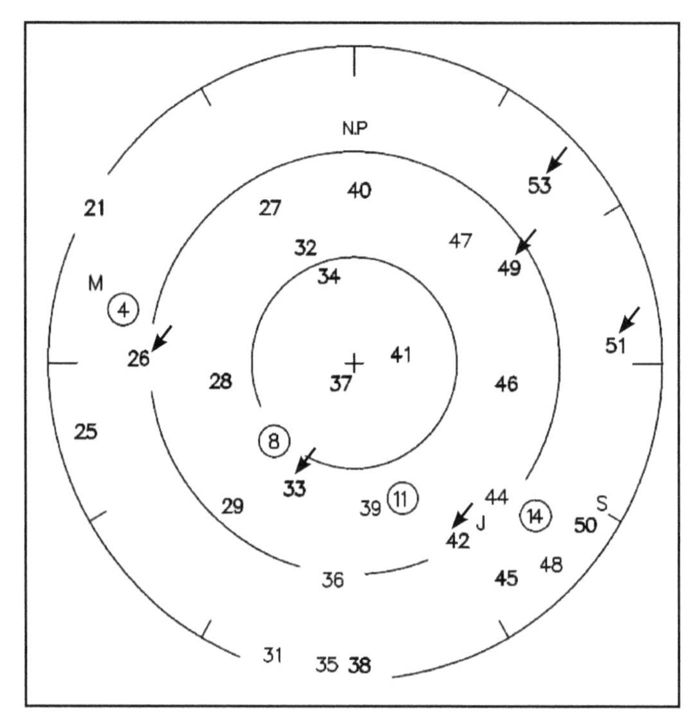

Figure A-3. *Sky for 25N, on July 15, 2019 at LMT 19h. Arrows mark the magnitude-1 stars. The planets M (Mars) and J (Jupiter) would also be bright.*

Body Selection Criteria for Most Accurate Position Fix

Condition 1. We want three bodies, as near as possible to 120° apart.

This is the primary criterion—there is no virtue in taking 4 stars, or even 10 stars. We want the three best, and then we want three or four sights of each one of them.

Condition 2. Bright stars should be favored.

Condition 3. If everything else is equal, the three should be as near the same height as possible.

Condition 4. Generally we want sights to be higher than about 10° and lower than about 75°.

Lower than 10° introduces refraction uncertainties; above 75° stresses the approximations we use to derive the lines of positions and sights higher than that are increasingly harder to take. Either one of these limits can be pushed some if a notably better 120° triad can be formed.

From a practical point, we can rule out the moon in this sky, because it is so low. Even if rather higher, it would not be a good choice, because it must be near a full moon, which is usually bright enough to distort the horizon below it (see our textbook previously mentioned). We know it is near full, because it is rising just after sunset. Mars is also on the limit of too low. Thus, as a first guess, we are looking for the best triad of the mag-1 stars, or stars and Jupiter.

In practice, we would print a Sky Diagram page from the AA, and highlight the bright stars. Then using a transparent sheet of some form, we draw on this transparency three intersecting bearing lines 120° apart. This we can do using any compass rose or universal plotting sheet. Then we overlay it on the diagram and rotate, with the centers aligned, to find the bodies nearest to 120° apart.

In Figure A-4 we use a plastic lid from a grocery store Chef's Salad as a good transparency sheet, overlaid on a Sky Diagram page printed on letter size paper, the default page size of the PDF. Such a transparent lid can also be used for depth sounding navigation, writing on it with a Vis-a-Vis marker, which dries but wipes clean with water. See *davidburchnavigation.blogspot.com/2014/01/ecs-without-gps.html* for an article that includes a section on line of soundings navigation and also this video at *passagemaker.com/technical/bathymetric-navigation-line-of-soundings-method* by navigation instructor and high-speed ferry operator Robert Reeder.

The Sky Diagrams are smaller than we would like, but still quite usable. I have to put it that way because the one-paragraph Section 2212 of the only Bowditch references to this method ('58 and '77) say the diagrams are of "limited value because of their small scale." We disagree. Figure A-5 is a detail of the work done on the printed sheet in Figure A-4. We rotate the lid to see which triads line up best with the green bearing lines drawn 120° apart.

To numerically compare best triads of sights, we need a way to weight the values of good separation (nearest 120°), good brightness, and good heights. If we call these relative weights 70%, 20% and 10%, with height limits of 10° to 75° and require stars brighter than mag-2.0, then the best triad is [51, 33, 32] by that

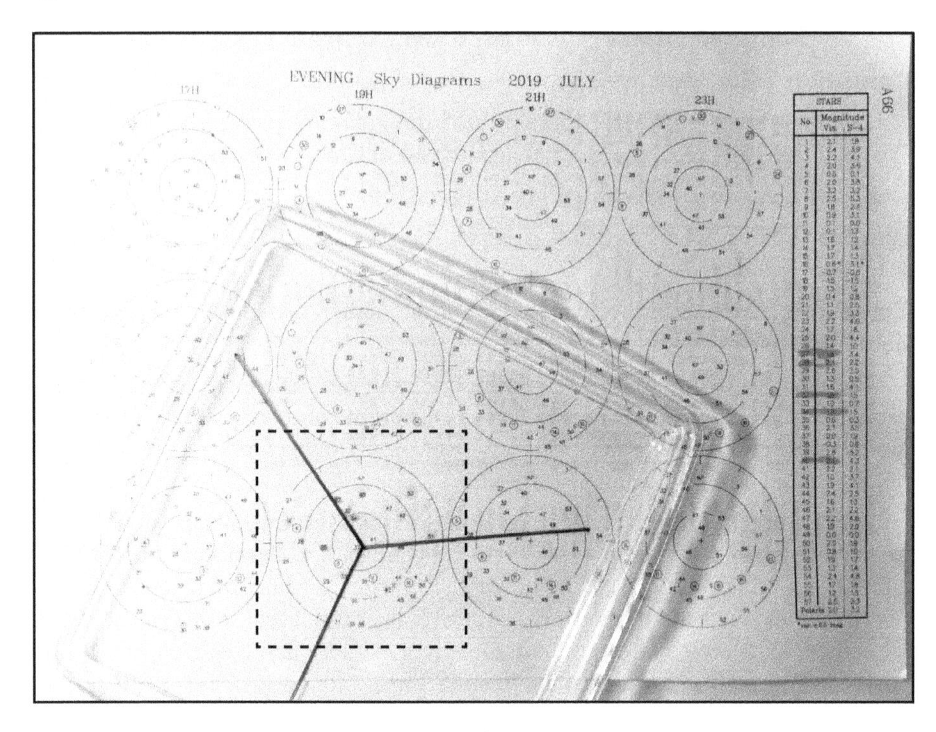

Figure A-4. *A full page, letter size print of the Sky Diagrams, with a transparent plastic sheet marked with 3 bearings, 120° apart. It can be rotated to find the 3 bodies that best align with this spacing. The dashed area is zoomed in Figure A-5, below.*

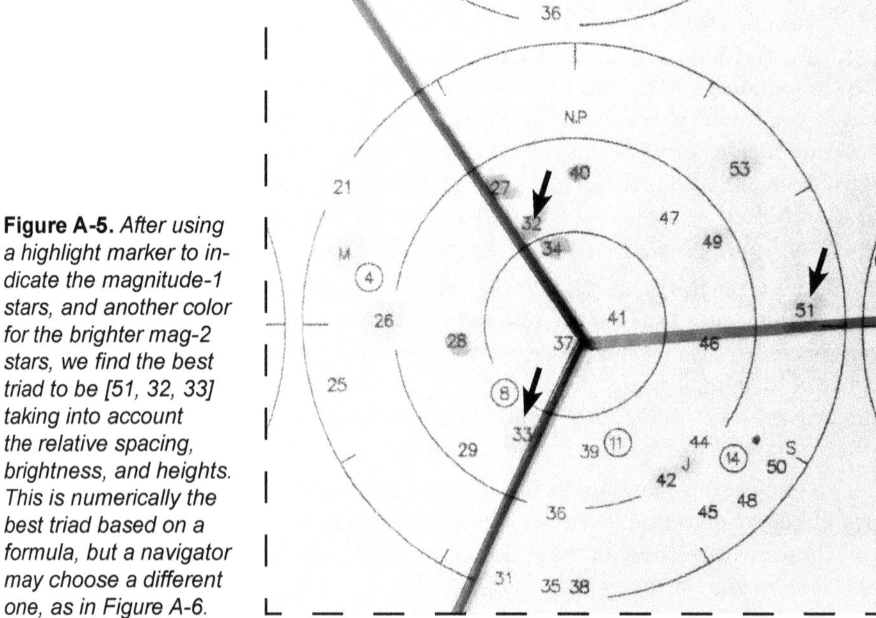

Figure A-5. *After using a highlight marker to indicate the magnitude-1 stars, and another color for the brighter mag-2 stars, we find the best triad to be [51, 32, 33] taking into account the relative spacing, brightness, and heights. This is numerically the best triad based on a formula, but a navigator may choose a different one, as in Figure A-6.*

standard—although it might not be the best in practice. It could be that one with just a slightly lower goodness factor would be preferred. We can get such a triad goodness factor from the StarPilot app's Find Best Sights function, which accounts for all these factors mathematically, looking at all possible triads that meet the user defined specs.

Computed values for these stars from the StarPilot output are:

Body	*Mag*	*Hs*	*Zn*
51:ALTAIR	0.8	013°08.3'	086.2°
33:SPICA	1.2	050°12.8'	206.3°
32:ALIOTH	1.8	054°18.4'	336.4°

The challenge in this sky is there are no bright objects in the NW sky. We have three mag-2 stars to choose from here: #27 at 1.8, #32 at 1.8, and #34 at 1.9. These are a toss up in brightness (magnitudes are listed to the right side of the diagrams page), but #32 wins out slightly in the equal heights factor, although that is the weakest of the conditions. In practice these are equivalent stars for that direction in this sky. On deck we would take the one that looked brightest or appealed to us for some other reason.

Figure A-6 is another contender where we favor brightness and give up a bit on the spacing factor. Here we choose [26, 42, 53], all mag-1 stars. These bright stars would make an attractive set, even though they are not best possible spacing, and technically not the optimum triad. The computed sight data are:

Body	*Mag*	*Hs*	*Zn*
53:DENEB	1.2	016°02.5'	046.9°
42:ANTARES	1.2	031°16.0'	149.2°
26:REGULUS	1.2	026°35.9'	271.2°

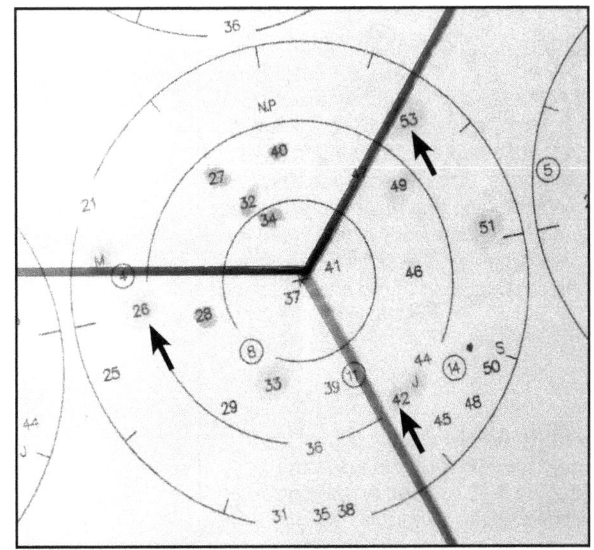

Figure A-6. *An optimization formula always takes the best answer even if just slightly better. The role of the navigator is to choose the triad that will be best in practice, and in this case, chances are this one with brighter stars might be it... but we have to remember that bright stars lower on the horizon will not be as bright as when higher up, so this one also has to be tested on deck.*

Pub 249 Vol. 1

For comparison to the Sky Diagrams, a popular way to select the best triad of stars is Pub 249 Vol. 1, a book issued every five years and good for eight. The latest is Epoch 2020. To use it we need a NA or AA to look up the GHA of Aries, then we subtract (or add in E Lon) our DR-Lon to get the LHA of Aries. For this type of star analysis we can simply assume our Lon is 0°, making the LHA = GHA and the LMT = UTC.

To match this exact Sky Diagram at LMT = 19h (recall we will do a random time and Lat later), we look up GHA Aries at UTC 19h on July 15 to get 218° 20.3'. We can then go to the page in Pub 249 for Lat = 25 N, and down to the LHA Aries we found.

As we see in Figure A-7, the conventions with this pub are the bright stars are in CAPS and the three best are marked with a diamond, namely [40, 42, 26] which have brightness 2.1. 1.0, and 1.4. The Pub 249 height limits on best sight choices are stricter than we might guess, being about 13° to 60°, which must be related to aircraft sextant sights from higher altitudes. A marine sextant sight at 75° is no more difficult nor uncertain than one at 60°. Also Pub 249 has more weight on sights of roughly equal height, which must be for the same reasons.

Figure A-8 shows the triad Pub 249 Vol. 1 proposes as we would discern it in a Sky Diagram. This would not be the first one we would choose from the many

	40 ♦Kochab	VEGA	Rasalhague	42 ♦ANTARES	SPICA	26 ♦REGULUS	Dubhe
210	40°28' 005	30°53' 058	37°45' 093	27°00' 142	52°49' 194	34°10' 268	43°04' 333
211	40 32 004	31 30 058	38 40 094	27 33 143	52 35 195	33 16 268	42 39 333
Pub 249 V1 LAT 25° N			39 34 094	28 05 144	52 20 197	32 21 268	42 14 333
			40 28 095	28 37 145	52 04 198	31 27 269	41 49 332
214	40 43 003	33 59 059	41 22 095	29 08 145	51 46 200	30 32 269	41 24 332
215	40°45' 003	34°45' 059	42°16' 096	29°39' 146	51°26' 201	29°38' 270	40°58' 332
216	40 48 002	35 32 059	43 10 096	30 09 147	51 06 203	28 44 270	40 32 332
217	40 50 002	36 18 059	44 04 097	30 38 148	50 44 204	27 49 271	40 06 331
218	40 52 002	37 05 059	44 58 097	31 06 149	50 21 206	26 55 271	39 40 331
219	40 53 001	37 51 059	45 52 098	31 34 150	49 57 207	26 01 271	39 13 331

Figure A-7. Above. *Section of Pub 249 Vol. 1, Epoch 2020 for Lat 25N. The box and star numbers were added. The triad proposed by Pub 240 Vol. 1 [40, 42, 26] is marked with diamonds. One advantage of this method, if the triad is accepted, is the precision of the Hc, which is used for presetting the sextant and thus allowing easy location of the stars.*

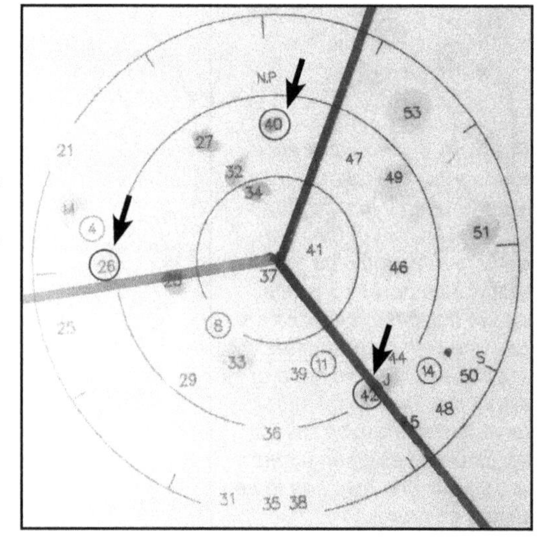

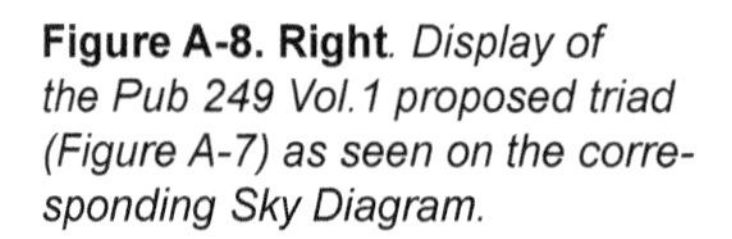

Figure A-8. Right. *Display of the Pub 249 Vol. 1 proposed triad (Figure A-7) as seen on the corresponding Sky Diagram.*

we can find with the Sky Diagrams. Pub 249 is a fast, convenient solution, but it does not always choose the optimum triad, and it can never find one that includes planets. We have to conclude that the Sky Diagrams are better than Pub 249 Vol. 1 for sight selection, faster and more complete—provided you invest in a good Chef's Salad lid or some equivalent.

2102-D Star Finder

Another popular way to predict best triads is the 2102-D Star Finder. This plastic device from the 1920s is still for sale, now as a commercial product, although in the 30s and 40s it was an official Navy Hydrographic Office issue, H.O. 2102-D. We are a strong supporter of this device (Figure A-9).

The device, with its various templates for different latitudes, is essentially a hand-held planetarium. You can see when and where bodies rise and set, find the best sights at twilight, or simply go on deck and take sights of three bodies in the right relative directions and not even worry who they are. With just a few minutes work you can with this device identify the stars you sighted. The separation of about 120° is found on deck by facing one chosen star, and then take a hard look over each shoulder. On the other hand, we can use it to predict the best three before taking the sights.

As with Pub 249 Vol. 1, we need an AA or NA to look up LHA Aries (218° in this example) and then, with the 25°N blue template in place on the white base plate, we rotate the template to align it with that LHA Aries value as shown in Figure A-10. The annotations we added show how to read heights (Hc) and bearings (Zn). The ones shown are for the [51, 32, 33] triad shown in Figure A-5.

In general, the way we do best sights selection with this device is to set it up with the right Lat (choice of blue template) and LHA (rotate template on white base plate), and then read off the Hc and Zn of the bright stars, which we then plot on a universal plotting sheet. A sample from *The Star Finder Book* is shown in Figure A-11 for a different sky.

In short, optimum star prediction with the Star Finder is just using it to make a big Sky Diagram! We could also annotate such a plot with star heights, but in this case we only plotted bodies with usable heights. To include the planets or moon, use the Star Finder's red template and other NA or AA data to plot them on the white base plate, then we can include them. But that is extra work that is all done for us with the Sky Diagrams.

The one advantage of the Star Finder solution is we are using exact times for the LHA setting, with latitude templates every

Figure A-9. *Cover of the third edition, 2019.*

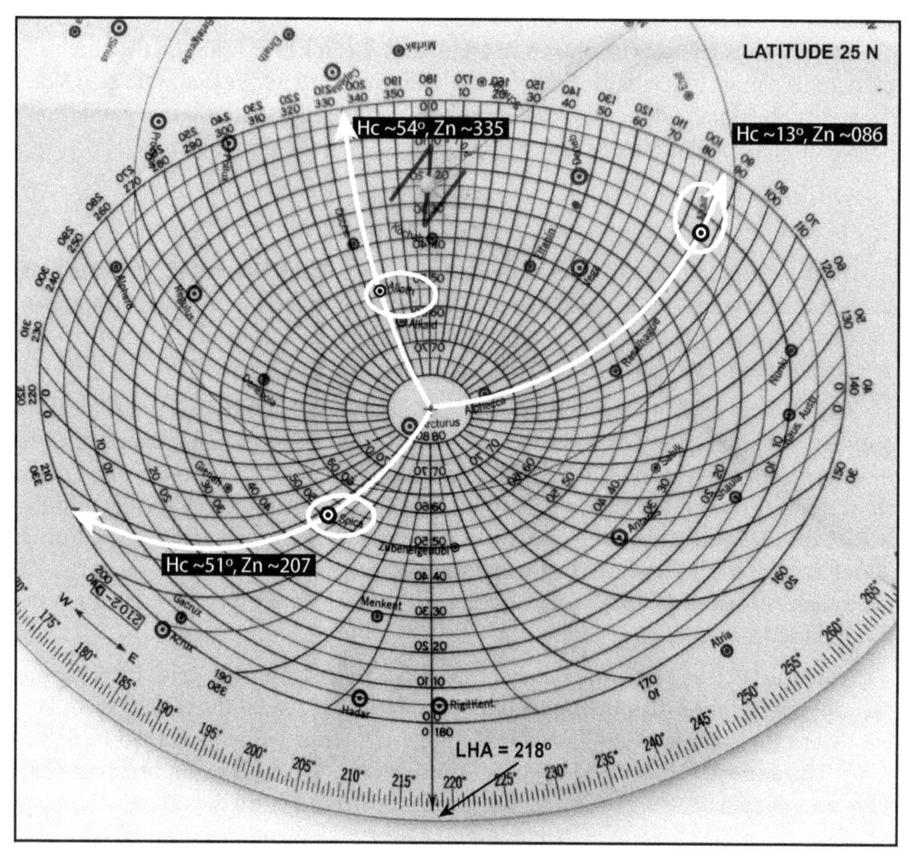

Figure A-10. *Star Finder set up for LHA = 218°, Lat = 25N. The triad [51, 32, 33] of stars are marked with how the height (Hc) and bearing (Zn) are read from the dials.*

10° rather than 25° in the Sky Diagrams, so we end up with more accurate values of the Hc and Zn, which are valuable for setting up the sextant to take the sights. Using Sky Diagrams the Hc and Zn we get are more approximate. On the other hand, once we have determined the best triad from the Sky Diagrams, we can compute accurate values of Hc and Zn for the actual sight taking using sight reduction tables.

Sky Diagrams at Random Sight Times and Latitudes

A long analysis with a simple conclusion

We have seen that in an idealized example, meaning a case that matches exactly the date, time, and Lat of one of the diagrams, the Sky Diagrams are an excellent, if not superior, way to manually find the best triad of sights. Now we need to see how much harder it is for realistic sight times that do not match one of the base diagrams. To do this, we chose a random but realistic evening twilight on July 10 at 31N, 140W. This involves, on some level, a triple interpolation: Lat, LMT, and Date.

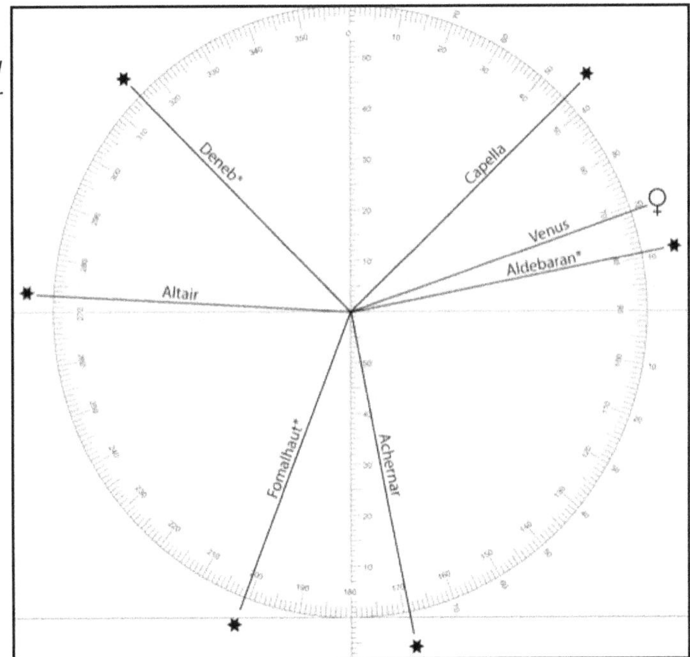

Figure A-11. *Plot of the potential sextant sights found from the Star Finder using a universal plotting sheet—for a different time and place than discussed. The heights and brightness could also be shown in this plot, but these targets have already been filtered for that; we are here just looking for the best spacing. A marked transparent sheet would assist with that selection.*

First we gather the real data, which we can get from NA or AA. Evening sights are taken between civil twilight and nautical twilight, which at 31N on July 10 (refer to our textbook cited earlier on this determination) are 1934 and 2007 LMT. We might then plan our sights for the midpoint of this 33-minute sight session, namely, 1950 LMT on July 10 at 31N.

We have base diagrams at 25N and 50N, so at 31N we are looking at a sky that is roughly halfway between these two. Our target time of 1950 is about 20h, which is halfway between the 19h and 21h given. Also we are looking at July 10, which is 5 days earlier than the base diagrams, which are all valid on the 15th of each month. To do this numerically seems daunting; the right sky is roughly the average of four of the base diagrams.

But do we really need to do that? After all, I know from direct experience that we can sail across the Pacific (WA to HI) and pretty much use the same three stars for most of the voyage, maybe shifting choices once, only toward the end. This is a latitude change 27° (48 to 21), spanning 3 time zone descriptions (+7 to +10). So chances are the top choice triads are not going to change much for rather large changes in latitude and date. So let's start by just rounding everything to the nearest diagram, which the Sky Diagram's explanation section hints is doable, referring to interpolation as "... such refinements are not usually necessary." But they are in airplanes and we are in boats, so we have to test this.

With that approach, a sight session spanning times of 1934 to 2007 would be rounded to 19h (there is no 20h diagram), and Lat 31N rounds to the 25N diagram, and for July 10 we use July 15 (the only date given), then we are back (coincidentally) to the example done above (July 15, 19h, 25N), whose technically best triad was [51, 33, 32]. We can now use StarPilot to see if this is the right answer for the actual conditions, 1950, July 10, at 31 N, which we compute specifically for those values, shown in Figure A-12.

And yes, indeed, that is the best triad. Not only does this rounding method work, in this case this triad at a special time and Lat is even better than the base diagram, as shown in this StarPilot output (Figure A-13).

Without going into details of the solution, we can see that even though the Hc and Zn of the stars are somewhat different in these two skies (the actual sky at hand and the nearest Sky Diagram), this is not just the same triad, it is numerically superior at the special time with a goodness weight of 2.2 compared to 1.8.

In short, we see here an easy way to test whether or not rounding everything to find the right Sky Diagram to use will be dependable: we just do that in multiple cases, and then compare the answers with the numerical solutions from StarPilot. The question is, how do we choose the random sight times to test? One way is we look at the vessel positions from any documented ocean crossing and choose a few positions from that. We just happen to have a perfect candidate: our published training exercise (*Hawaii by Sextant*) based on a July, 1982 voyage from Cape Flattery, WA to Maui, HI, carried out purely by cel nav. We will just assume this took place in July, 2019 and select several sight times and locations along that hypothetical voyage. Figure A-14 shows the route and the points we choose to test.

First the easy part, we look at what triads the StarPilot found for evening and morning twilights at the positions and dates shown in Figure A-14. The optimum triad solutions are computed using the exact twilight times and latitudes of these hypothetical sight sessions. The results are shown in Figure A-15. The Lat-Lon

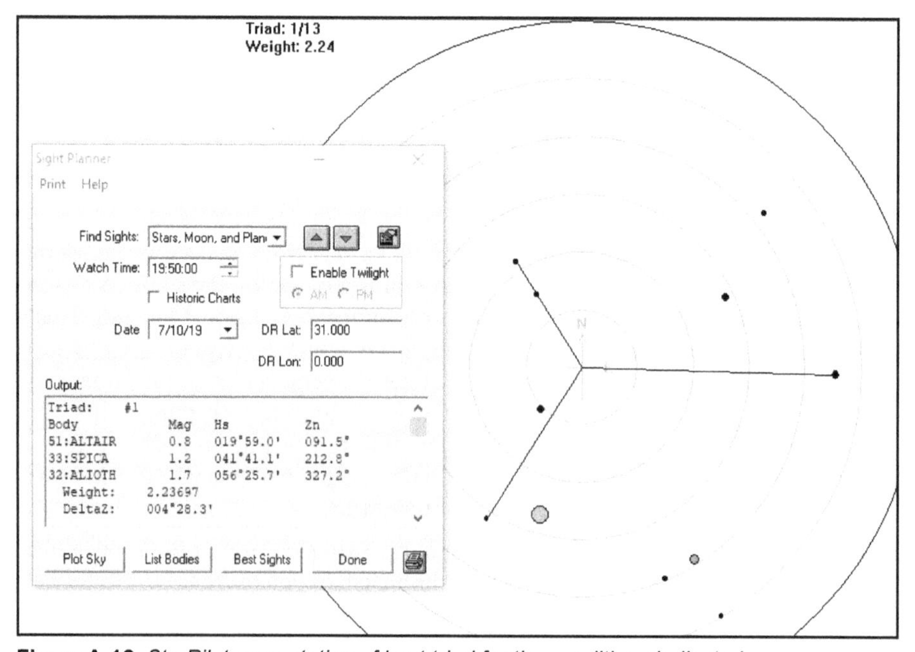

Figure A-12. *StarPilot computation of best triad for the conditions indicated.*

Base AA Sky Diagram

Triad: #1 July 15, Lat 25N, 1900 LMT

Body	Mag	Hs	Zn
51:ALTAIR	0.8	013°08.3'	086.2°
33:SPICA	1.2	050°12.8'	206.3°
32:ALIOTH	1.7	054°18.4'	336.4°

Weight: 1.80551
DeltaZ: 010°12.3'
DeltaH: 001°14.0'
DeltaI: 3.2

Special case

Triad: #1 July 10, Lat 31N, 1950 LMT

Body	Mag	Hs	Zn
51:ALTAIR	0.8	019°59.0'	091.5°
33:SPICA	1.2	041°41.1'	212.8°
32:ALIOTH	1.7	056°25.7'	327.2°

Weight: 2.23697
DeltaZ: 004°28.3'
DeltaH: 001°14.0'
DeltaI: 2.6

Figure A-13. *Comparison of best computed triad for a specific time and place (right) with best triad determined from the Sky Diagrams by rounding all values to one of the tabulated base diagrams (left), showing that this severe rounding still leads to the best choice.*

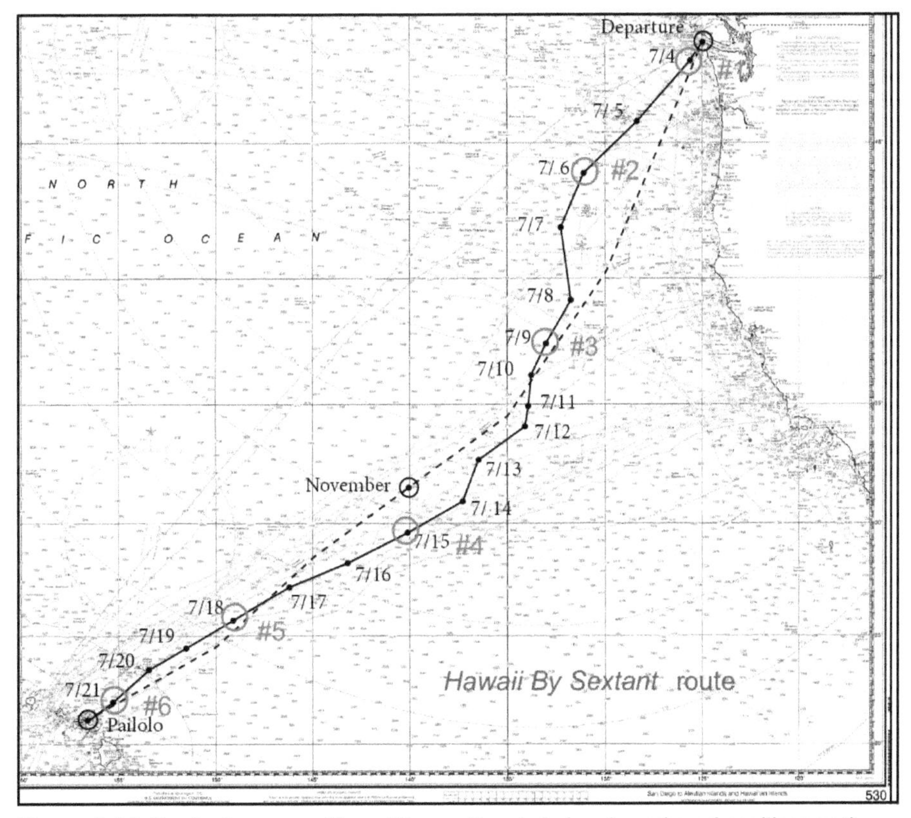

Figure A-14. *Track of noon positions. We use the circled and numbered positions and dates for the year 2019 to test the Sky Diagrams prediction of best triads.*

values shown are in the StarPilot's shorthand format, designed for quick input: 34.567 means 34° 56.7', etc.

Next we take these real conditions and round everything to a base diagram. The rounded skies are shown in Figure A-16. Now we ask StarPilot to find the best triad in these rounded skies and compare those results with the choices made for the actual skies before rounding. Figure A-17 shows the morning sights. The dashed-line triads are the ones we know from the exact data are the technically optimum choices. The solid-line triads are the ones we would have selected from these base diagrams using the same quality criteria. We get the right choices, even in this dramatic rounding of all values, in all cases but #6 and this one is very close.

Figure A-18 is the evening sights, with the same conventions: dashed lines are the known optimum triads for the precise conditions, and solid lines are what we would have chosen from these base diagrams without knowing anything more. In all cases but evening sights #3, we get the right choice by total rounding to the base charts available.

However, we need to look closer to see if #3 is really wrong. The actual Lat for this session was 37° 44', which is just barely leaning to 50N over 25N. In fact, we have several arguments here to override the strict Lat rounding, and instead go to

StarPilot Best Sights Solutions							
#	Date	Lat	Lon	Morning LMT	Triad 1	Evening LMT	Triad 1
1	4-Jul	46.09	127.58	0245	12, 56, 49	2032	53, 42, 32
2	6-Jul	43.47	131.12	0303	12, 56, 49	2020	53, 42, 34
3	9-Jul	37.44	132.31	0337	9, 56, 49	1955	51, 37, 27
4	15-Jul	29.48	139.53	0410	14, 56, 49	1929	51, 33, 32
5	18-Jul	25.55	149.09	0423	14, 56, 49	1919	51, 33, 32
6	21-Jul	22.32	154.31	0432	10, 56, 53	1910	51, 33, 32

Figure A-15. *Input data used with the StarPilot program to determine the best triads for each position. Shorthand notation in use: 37.44 = 37° 44'. These results represent the actual conditions. They are plotted as dashed lines in Figures A-17 and A-18.*

Sights Rounded to Base Diagrams						
#	Date	Lat (25 or 50)	Morning LMT (1h, 3h, 5h,7h)	Best Morning Triad	Evening LMT (17h, 19h, 21h, 23h)	Best Evening Triad
1	15-Jul	50	3h	12, 56, 49	21h	53, 42, 32
2	15-Jul	50	3h	12, 56, 49	21h	53, 42, 32
3	15-Jul	50	3h	12, 56, 49	19h	12, 49, 33
4	15-Jul	25	5h	14, 56, 49	19h	51, 33, 32
5	15-Jul	25	5h	14, 56, 49	19h	51, 33, 32
6	15-Jul	25	5h	14, 56, 49	19h	51, 33, 32

Figure A-16. *Optimum triads found from the skies obtained by rounding the sights of A-15 to nearest base diagram. They are plotted as solid lines in Figures A-17 and A-18.*

25N. From the diagram we see that at 50N the sun is still up, and by the time it sets, the chosen #12 (bright Capella) would not be an option. So it would not take much reasoning to use 25N, and in that case all matches up.

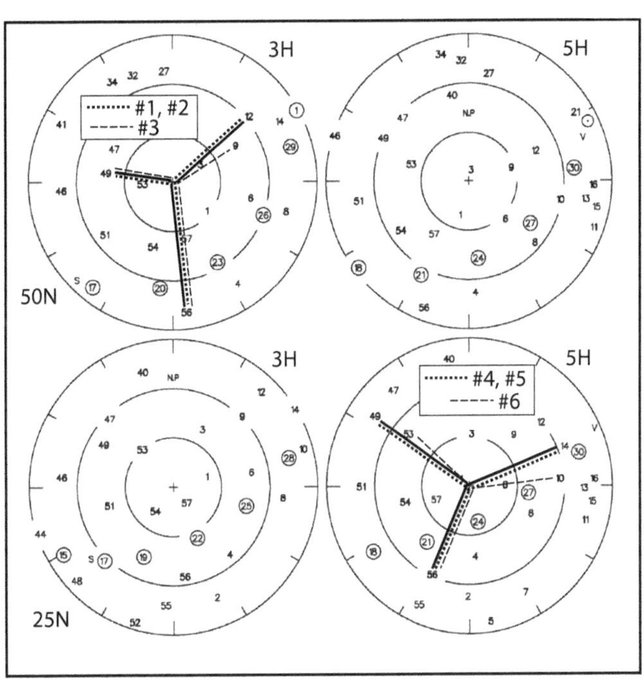

Figure A-17. *Morning sights. Dashed lines are those found for the actual sight times of Figure A-15; solid lines are the best triads of the rounded sights of Figure A-16.*

The solid lines agree with the dashed lines (other than two minor discrepancies) showing that the best triads can be found by rounding to the base diagrams.

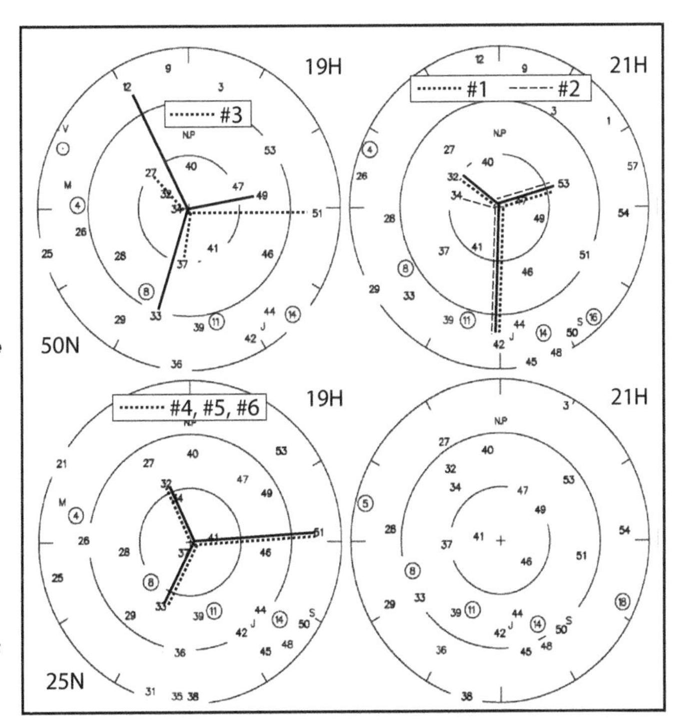

Figure A-18. *Evening sights. Dashed lines are those found for the actual sight times of Figure A-15; solid lines are the best triads of the rounded sights of Figure A-16.*

The rounded sights agreed except for #2 slightly off and #3 notably off, but there is an explanation for #3 covered in the text.

I think that is all the testing we need to do! Recall, too, that these are analyses and decisions we make for the first sight session in a cel nav voyage. If we choose something wrong the first night, we then know it, and how to correct it. Then we are right the rest of the voyage until this quick look at the Sky Diagrams changes our mind.

Also I should point out that all of this best bodies planning can be done before you leave the dock! We see that there is broad leeway in times and latitudes that still come to the right choices, so you can DR your crossing and make a table of best bodies for each region of the route. Chances are you will have the right choices in hand as you cross without further work. It is an excellent exercise that gets you thinking about the sky before you get there, which is just one more step along the path of good seamanship, which is preparation. For practice with this you can use the actual route in *Hawaii by Sextant*, and from the support page of that book (starpath.com/HBS) download the 1982 Sky Diagrams, and compare what the options were with actual sights taken.

A bonus of the diagrams is, once we find our target triad, we can draw in our heading line, as shown in Figure A-19, and this way quickly picture where the stars are located relative to the bow. In this example, we have bright ones on the port bow and quarter, and the fainter one roughly on the starboard beam. With such a layout you can anticipate if it will be needed to change headings, if possible, during a sight session, or maybe set sails will prevent this triad, so you look for another.

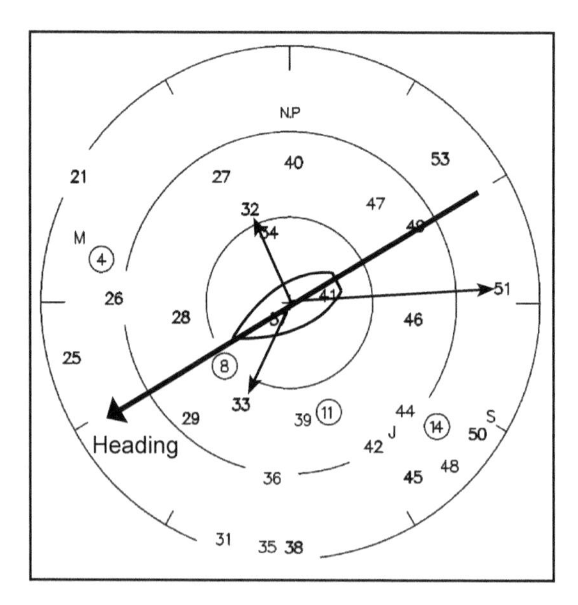

Figure A-19. *A vessel heading line drawn on a Sky Diagram to determine relative bearings of the target bodies.*

Star Brightness and Sun-Moon Daytime Fix

There are two tables, taken from *The Star Finder Book,* that help with choosing best sights. Table A-1 tells how perceived brightness differences can be determined from magnitude differences. A mag 1.7 star, for example, is about 40% brighter than a mag 2.1 star. Pub 249 Vol. 1 with its height limit of 60° often has to dip into the mag-2 stars, which sometimes limits its ability to find the best triad.

Table A-2 tells what days and times of month we can get a good sun-moon fix during the day, along with other special uses of the moon. In Table A-2 we learn that one chance for a good sun-moon fix is moon age days 21-23, a waning half moon near the meridian at sunrise, with best sight time in the mid morning. In

Table A-1. Brightness vs. Magnitude

Magnitude Difference	*Brightness Difference*	*Magnitude Difference*	*Brightness Difference*	*Magnitude Difference*	*Brightness Difference*
0.0	1.0	2.0	6.3	4.0	40
0.2	1.2	2.2	7.6	4.5	63
0.4	1.4	2.4	9.1	5.0	100
0.6	1.7	2.6	11	5.5	158
0.8	2.1	2.8	13	6.0	251
1.0	2.5	3.0	16	6.5	398
1.2	3.0	3.2	19		
1.4	3.6	3.4	23		
1.6	4.4	3.6	28		
1.8	5.2	3.8	33		

Table A-2. Guidelines for the Use of the Moon

Age (days)	Phase and Location	Sight Time	Special Value
1-3	waxing crescent, setting near western horizon at sunset	just after sunset, before evening stars appear	combine with evening star sights
6-8	waxing half moon, near the meridian sunset	mid afternoon	combine with sun at for day-time fix
12-14	waxing full moon, rising near eastern horizon at sunset	just after sunset, before evening stars appear	combine with evening star sights
15-17	waning full moon, setting near western horizon at sunrise	just before sunrise, after morning stars fade	combine with morning star sights
21-23	waning half moon, near the meridian at sunrise	mid morning	combine with sun for daytime fix
26-28	waning crescent, rising near eastern horizon at sunrise	just before sunrise, after morning stars fade	combine with morning star sights

Table Notes: *The range of ages might be off a day or so in some circumstances. The predictions are least dependable when either sun or moon passes near overhead during the days in question, meaning best use of the moon depends on your latitude and the declination of the moon. Moon age is on the daily pages of the* Nautical Almanac.

principle we can also get such information from the Daytime Skies of the Sky Diagrams. We want the sun and moon about 90° apart, which means a half moon. The days of the half moons can be found from the NA, which lists moon age on the daily pages (look for age 6 to 8 or 21 to 23), plus there is a table of moon phases on page 4, or from the AA, which lists illumination on the daily pages (look for about 50%). In the latter we can also find these dates at a glance from the Semi-duration of Moonlight Table, intended for polar navigation.

From the Semi-duration table, in July, 2019 (page A155) we have a waxing half moon on day 9, and a waning one on day 25, which can be confirmed with their illumination data. Now we can note these positions on the daytime sky diagrams to mark when we should look out for good sun-moon fixes (Figure A-20). The best sun-moon fix will not necessarily be on that day, but could be a day or so before or after, which is why we need to study this.

Figure A-20 shows only one time (9H) but this can be shown on any of the daytime diagrams for waxing or waning half moons. We see that the exact days of half moon are not the optimum sight days in this case, which we find from the Sky Diagrams by making their relative bearings near 90°. We also learn here at home what many celestial navigators once learned underway. Namely, sailing into the tropics in late June or early July leads to difficult sun only navigation. Sailing under the sun, those relying on noon sights alone are stuck for some days without sights. The sun is just too high at midday. Thus, we recommend learning all cel nav, not just sun sights.

Here we see that at latitudes anywhere near the sun or moon's declination, sun-moon sights will be very difficult. One of them will be near overhead. These sights are difficult because the crucial technique of rocking the sextant is difficult, because we do not know which way to look when we rock. Our textbook discusses the ways around this.

On the other hand, at higher or lower latitudes we can spot the best times and days do to the sights with the Sky Diagrams. In this application, we are using the plotted moon positions to tell us the date, and then we assume that the sun's position on that date at that time will be about where it is on the 15th.

Figure A-21 shows sample results of testing these conclusions with the StarPilot. Namely, we take the proposed time and date based on the exact time of half moon, and then we shift the date to improve the sights based on Sky Diagrams, then plot out this proposed sky with StarPilot.

Only two are presented, but all show good 90° intersections. In practice we have a 2 or 3 day window on these sights for usable fixes, but this way we can spot the best time. Indeed, with a highlight marker for sun and moon we do not even need to look into half moon times. We can go direct to the diagrams and look for 90° opportunities. Likewise for sun-Venus sights. The Sky Diagrams do a fine job on these daylight sights.

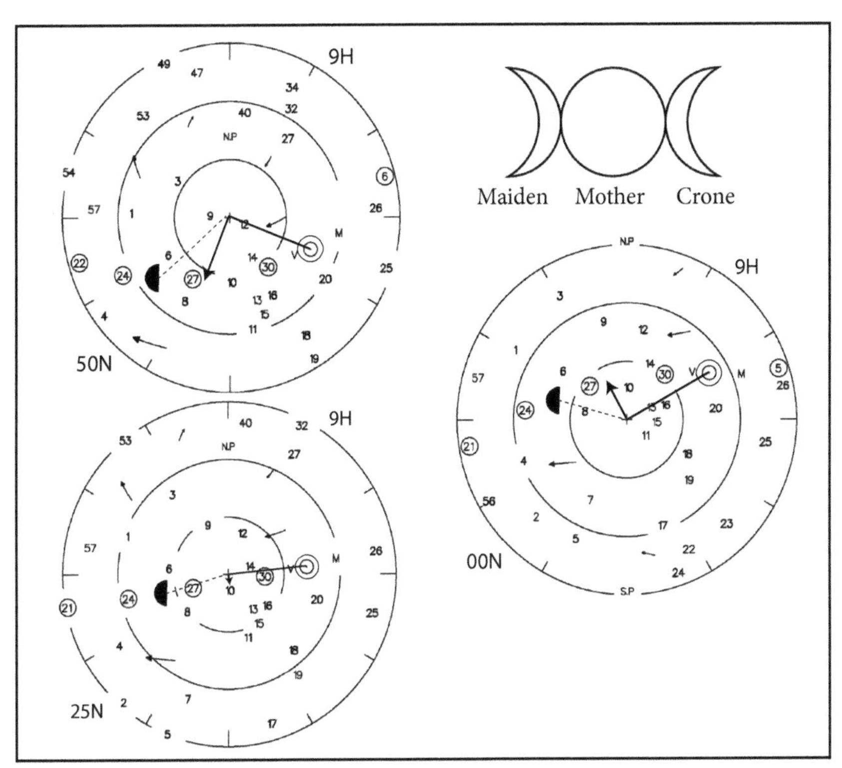

Figure A-20. *Location of the sun and moon at three latitudes as a function of LMT and date. Locations of the precise dates of a half moon (25th) are shown with a half-moon symbol. We learn, for example, that at 50 N, we get a better sun-moon intersection (i.e., closer to 90°) on the 27th or 28th rather than on the exact half-moon date. We also see that even though the phase of the moon is right, at latitudes near 25 N, the sun-moon angles and heights rule out a good sun-moon fix. The Celtic Goddess symbol is a reminder of the phase. See discussion at* davidburchnavigation.blogspot.com/2016/05/is-moon-waning-or-waxing.html

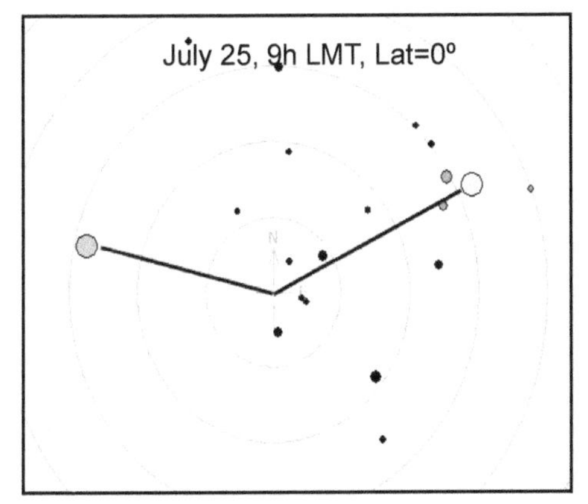

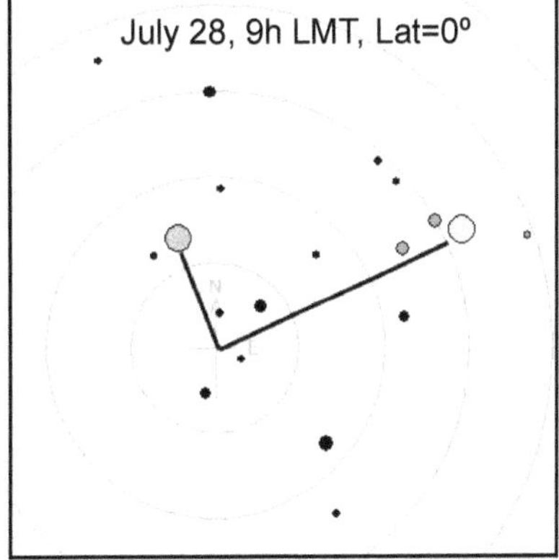

Figure A-21. *StarPilot computations of the sky for 9H on at the equator on July 25 (left), the day of the full moon, and (right) showing that the intersections are better on July 28, something we can learn from the Sky Diagrams.*

Grand Summary on the Sky Diagrams

My conclusion is that a printed copy of the Air Almanac's annual Sky Diagrams and a transparent plastic container lid is the easiest and fastest way to manually choose the optimum triad of celestial bodies for a sextant sight session. Having several colors of highlight markers helps. The diagrams are 76 pages of a free PDF copy of the Air Almanac, published annually. We recommend they be considered part of the celestial navigator's standard toolkit.

I had to specify "manual" solution in that conclusion, because if we slip into the world of electronics, we get other solutions, notably the StarPilot, which offers state of the art sight planning. That process is discussed in the StarPilot User's Guides, and related videos. Its quick and easy interface was key to the detailed analysis carried out above.

The attraction of the manual solution is that those studying cel nav, if not to pass a USCG exam, are generally doing it to be a backup to electronics failures on an ocean passage. In that event, we go back to old school cel nav: watches, books, and plotting sheets. What we are suggesting here is that a print-out of the few pages of Sky Diagrams that cover your voyage would be a good addition to your preparation of for an all-manual solution. Not to mention that the full (free) PDF of the Air Almanac is a backup to your other almanac data.

Let me conclude with reference to a discussion of the broader question: Why study celestial navigation at all in the age of GPS? *davidburchnavigation.blogspot.com/2013/10/why-study-celestial-navigation-in-age.html*

Notes on Star Motions in the Sky Diagrams —an Appendix to the Appendix.

I hope it is clear from the above notes that we do not need to know more about these diagrams to efficiently use them to find the best triads to start with in any sextant sight session. But there are ways presented in the official explanations of the diagrams to fine tune the interpolations based on known star motions and how they appear in these diagrams. Here we take a look at this issue, and further test if we need to consider these methods.

Star locations within each diagram are affected by two major astronomical motions. First, the earth circles the sun once a year (360° in 365 days), which is about 1° per day. This means that at the same time on consecutive evenings, the stars will be about 1° farther to the west than they were the previous evening at that time. This is indeed how the summer sky evolves into the winter sky—how the first sighting of Sirius before sunrise told ancient Egyptians the Nile was about to flood.

This orbital star motion is indicated by short arrows on the left-hand diagrams of each page (1h, 9h, 17h). A sample is shown in Figure A-22 that we have annotated.

Each diagram is valid for the 15th of the month. These arrows represent how far, and in what direction, a star in that vicinity will move during the 15 days following the 15th of the month, and from this we can project back to show where it would have been 15 days prior to the 15th. This motion is what we use to estimate star positions for dates other than the 15th. This is a date, or orbital correction.

The rings about the pole that we added are fairly good star tracks on this diagram at high latitudes, but these paths are squashed into ellipses at lower latitudes.

The second motion we need to account for is a time-of-day or rotation correction to account for the rotation of the earth about its axis on any given day. This causes the stars to move 360° in 24 hr, or 15° per hour, which is the difference between, say, the 17h diagram and the 19h diagram. We can see that in the diagrams themselves.

To illustrate this daily motion, we show in Figure A-23, the 21h diagram overlaid with the 17h and 19h locations of stars #53 and #46. At 15° per hour, in two hours they move about 30°.

Despite the complexity of depicting a dome of stars as a flat sky diagram, these two motions lead to what many navigators are well aware of. Namely, the sky we see above us now will be exactly what a navigator located 15° of Lon west of us will see one hour later. The dome of fixed star positions rotates about the pole, 15° each hour. This does not apply to moon or planet positions that are notably moving relative to the stars.

Perhaps less often thought of, the sky a navigator sees on the 15th of the month at a specific time is the same that was seen an hour later from the same place on the first of the month. To see how this works, rotate the sky backwards by 1° per day for 15 days, and then advance the time we look at it by 1 hr. The 1-hr forward in daily rotation of 15° clockwise, just cancels out the 15 days of backward orbital correction of 1° per day counterclockwise. Likewise, the sky we see on the 15th of the month is the same as will be seen on the 30th of the month, one hour earlier.

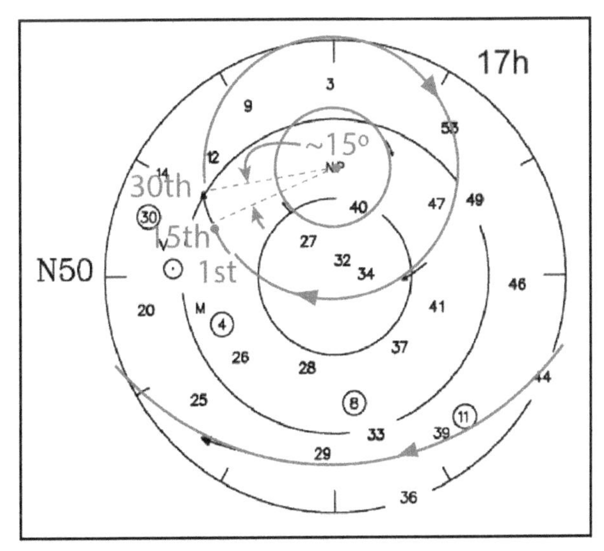

Figure A-22. *The daily motion of stars due to the yearly orbit of the earth around the sun.*

In other words, the 19h diagram that is valid on the 15th of the month is also valid on the 1st of the month at 20h, and on the 30th of the month at 18h. Figure A-24 shows a couple screen caps from the StarPilot app to illustrate this behavior at 25° N.

On the 15th at 19h, we see the stars with an optimum triad marked, plus the moon, Jupiter, and Mars. The moon was not visible in the other skies, and we see Mars notably move relative to the stars in these views. The star motion rules we are discussing apply only to the stars. In this case, Jupiter must be so far away from the sun that it effectively behaves like a star. This is common for Jupiter and Saturn, but Mars and Venus will usually move notably from day to day.

The summary of this is: Back 15 days and forward 1 hr gets to the same sky; Forward 15 days and back 1 hr gets to the same sky of stars—moon and planets not counted. Put another way, we can round the actual sight date not just to the 15th, but we can use the 1st or 30th, whichever is closer. If the 1st is closer than the 15th, then use the diagram that is 1 hr later than your sight time, and if the 30th is the closest, use the diagram that is 1 hr earlier.

So with that background, we see that we could fine tune the choice of diagrams depending on the date relative to the 15th—but we have also seen that this is not necessary!

We include this discussion of star motion because it is mentioned in the official explanation to the diagrams, but without details. To that I must add that I have in fact looked for cases where we could take advantage of this date correction for better choices, and I could not find any. In about half the cases, the date-corrected time will be an even hour, for which there is no diagram, so we are left interpolating again.

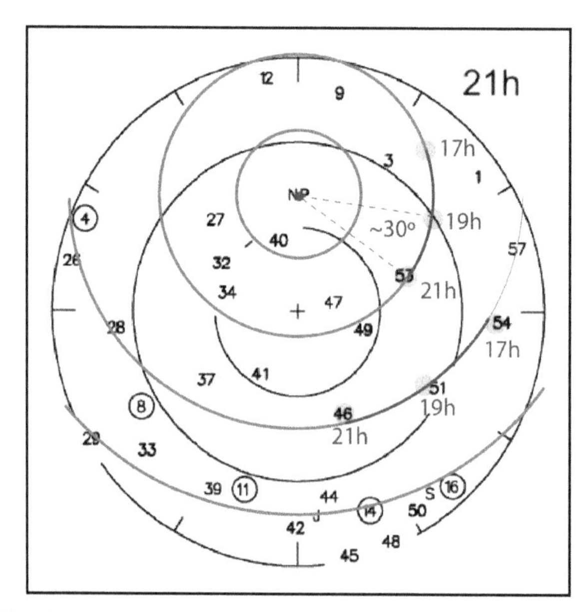

Figure A-23. *Hourly motion of the stars due to the earth's daily rotation on its axis.*

The earlier conclusion stands: there is no need for further interpolation. Round everything, and choose your stars. If near halfway between any two base latitudes (0, 25, 50, 75) then look for and consider other factors that might cause you to lean one way or the other.

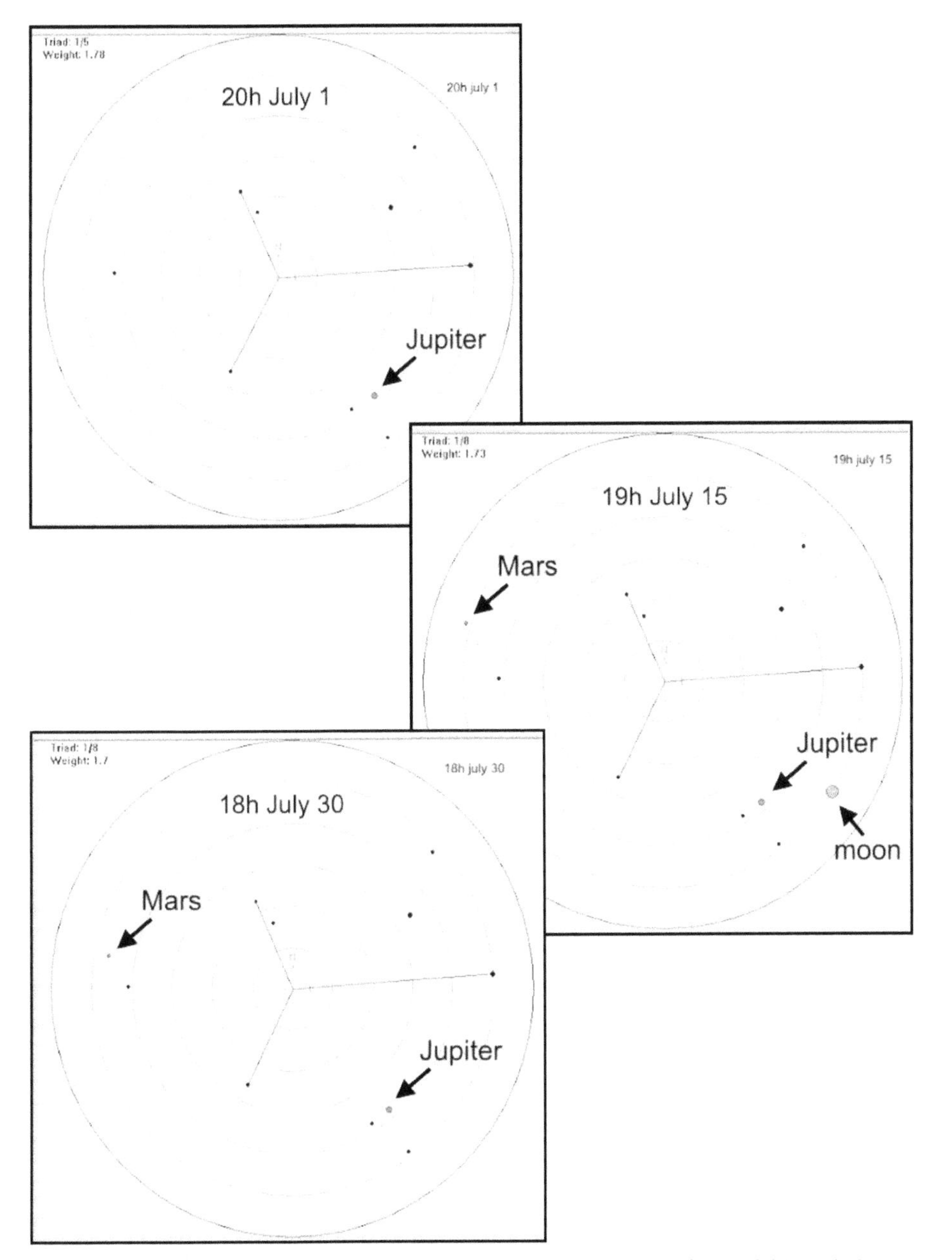

Figure A-24. *StarPilot computed skies showing: Back 15 days and forward 1 hr gets to the same sky; Forward 15 days and back 1 hr gets to the same sky of stars—moon and planets not counted.*

A2. Navigational Star Chart and Sky Diagrams

A copy of the AA's Navigational Star Chart is shown earlier in Figure 4-2 and in that caption there is a prescription on how to find the place overhead at any time, namely, the point on the star chart overhead of a specific Lat and Lon is given by:

Declination = Latitude

SHA = Lon W – GHA Aries or SHA = (360 – Lon E) – GHA Aries,

with a note that if SHA is negative, then SHA = SHA + 360°. In this section we provide an example of that process and compare it to what we get from a Sky Diagram covering the same sky.

We use an example of 19h (LMT) on July 15, 2019 at 25°N, 0°W, which makes the time = 1900 UTC. The SHA on the map is then 0° – 218.3° (+360) = 141.7°. This point is plotted on the chart in Figure A-26, and Figure A-25 is the corresponding Sky Diagram.

In the Sky Diagram the center is overhead and the horizon is 90° away on the circumference. This range is sketched in on the chart, noting the horizontal and vertical scales are different. Also the star chart SHA scale is backwards relative to the clockwise rotation of stars on the Sky Diagram, so in the star chart, east and west are reversed. It is as if we are under it looking up.

It is not clear how useful this is for actual start ID, but it is a way to estimate the sky from the chart alone. It also further illustrates the great virtue of the Sky Diagrams.

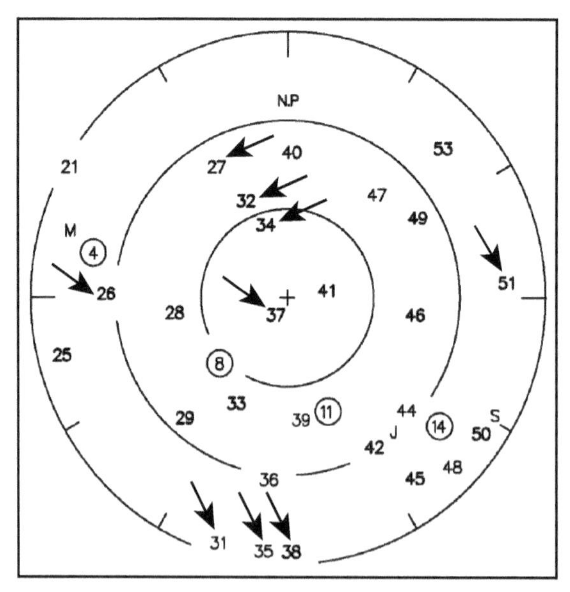

Figure A-25. *The base Sky Diagram for 25N, 19h LMT, July 15, which we compare to navigational star chart centered over that position in Figure A-25. Arrows mark the stars circled in Figure A-25.*

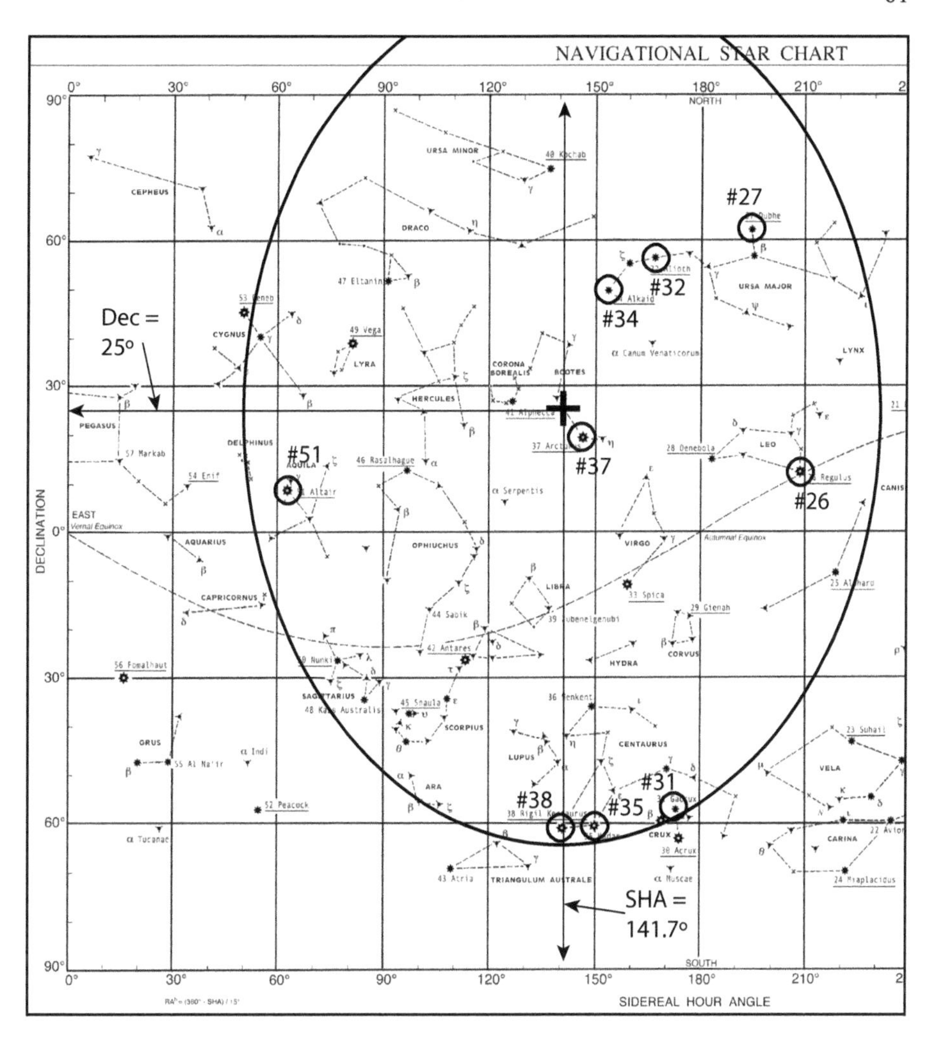

Figure A-26. *Section of the navigational star chart from the AA centered overhead of 25ºN, 0ºW on July 15, 2019. The solid line is a rough sketch of the horizon located 90º from the overhead point. The circled stars are marked by arrows in the Sky Diagram of Figure A-25. These star positions and bearings are distorted in the star chart, but shown to scale in the Sky Diagrams. Note that the star chart is perpetual, whereas the Sky Diagrams apply to a specific month and year.*

Other Starpath Publications

See starpath.com for more details.

CELESTIAL NAVIGATION

- *Celestial Navigation: A complete Home Study Course*
- *Hawaii by Sextant: An In-depth Exercise in Celestial Navigation Using Real Sextant Sights and Logbook Entries.*
- *How to use Plastic Sextants: With Applications to Metal Sextants and a Review of Sextant Piloting*
- *Starpath Celestial Navigation Work Forms: For All Sights and Tables, with Complete Instructions and Examples*
- *GPS Backup with a Mark 3 Sextant: All Instructions and Tables Included; For Any Ocean on Any Date; No Background in Celestial Navigation Required.*
- *Emergency Navigation: Find Your Position and Shape Your Course at Sea Even If Your Instruments Fail*
- Emergency Navigation plastic card: Tips and Tricks to Find Your Way on Land or Sea
- *Long Term Almanac 2000 to 2050: For the Sun and Selected Stars, with a Copy of the NAO Sight Reduction Tables*
- *Stark Tables: For Clearing the Lunar Distance*
- Starpath StarPilot series: State of the Art in Piloting and Celestial Navigation Computations (StarPilotLLC.com)

INLAND AND COASTAL NAVIGATION

- *Inland and Coastal Navigation: For Power-Driven and Sailing Vessels*
- *Navigation Workbook 18465 Tr*
- *Navigation Workbook 1210 Tr*
- *Navigation Workbook For Practice Underway*
- *Introduction to Electronic Chart Navigation: With an Annotated ECDIS Chart No. 1*

MARINE WEATHER

- *Modern Marine Weather: From Time-Honored Traditional Knowledge to the Latest Technologies*
- *Weather Workbook: Questions, Answers, and Resources on Marine Weather*
- *The Barometer Handbook: A Modern Look at Barometers and Applications of atmospheric Pressure*
- *Mariner's Pressure Atlas*
- Starpath Weather Trainer (starpath.com/wx)

CPSIA information can be obtained
at www.ICGtesting.com
Printed in the USA
LVHW050744231219
641446LV00008B/754